The Economics of Industrial Innovation

Second Edition

Christopher Freeman

1982

 The MIT Press, Cambridge, Massachusetts

Acknowledgements

An earlier version of this book was published by Penguin Modern Economic Texts in 1974. For this edition it has been extensively revised to take account of new research findings in the 1970s and early 1980s. A new chapter has also been added to deal with the issue of technical innovation and unemployment, which has come to the forefront of public attention in the last decade. This problem has been dealt with much more extensively in another book in which I collaborated with two colleagues (Freeman, Clark and Soete, 1982). I am grateful to them and to the publishers (Frances Pinter) for permission to use some excerpts from this book. The book also makes use of some papers which were previously published in various journals and reports. I am grateful to the National Institute of Economic and Social Research for permission to use excerpts from Numbers 26 and 34 of the *National Institute Economic Review* in chapters 3 and 4; to the Department of Trade and Industry for permission to use excerpts from a paper given to the Conference on Monopolies, Mergers and Restrictive Practices in chapter 6 (Ed. Heath, 1971); to the International Economic Association for permission to use excerpts from a paper given to their Conference on innovation (St Anton, 1971) in chapter 5; and to *Science Studies* for permission to use excerpts from a paper (Vol. 1, No. 3) in chapter 9. I have made numerous changes from the original versions, for which, of course, these organizations and journals bear no responsibility. Tables 2.1, 2.2 and 2.3 are reprinted from *Petroleum Progress and Profits*, by J. L. Enos, by permission of the MIT Press, copyright © 1962 by the Massachusetts Institute of Technology.

In writing the book I have been conscious all the time of my debt to colleagues who have worked with me at the Science Policy Research Unit and who have assisted me in a great variety of ways. Much of their work is cited, but I am particularly grateful to Keith Pavitt, Roy Rothwell and Joe Townsend for their help and advice over many years on questions of innovation. I am also grateful to colleagues who worked with me in earlier days at the National Institute of Economic and Social Research, especially Tibor Barna and Christopher Saunders, and to colleagues at the OECD, particularly Yvan Fabian and Alison Young, for their help with R & D statistics. The OECD statistics for which they have been responsible have been an invaluable help for all researchers in this field, and I am grateful to the OECD for permission to use these statistics and excerpts from their (Frascati) manual on the measurement of scientific and technical activities in the Appendix. International collaboration in this field has always been important and I have also benefited a great deal from the work and the advice of many fellow researchers in all parts of the world and especially those who have come to work for various periods at SPRU. I am particularly grateful in this connection to Geoff Oldham, a friend and colleague since the inception of SPRU and its present Director, who has done more than anyone to foster this spirit of internationalism.

Since the Science Policy Research Unit was established in 1966, many similar groups have been formed in all parts of the world and the interchange of experience has been extremely helpful, as for example with Herb Holloman, Jim Utterback and their colleagues in the Centre for Policy Alternatives at MIT or with Bengt-Åke Lundvall and his colleagues at the Aalborg Institute for Production in Denmark. It would be impossible to do justice to all the many individuals from whose work I have benefited, both from their published work and through their

1   By Way of an Introduction

In the world of microelectronics and genetic engineering, it is unnecessary to belabour the importance of science and technology for the economy. Whether like the sociologist, Marcuse, or the novelist, Simone de Beauvoir, we see technology primarily as a means of human enslavement and destruction, or whether, like Adam Smith and Marx, we see it primarily as a liberating force, we are all involved in its advance. However much we might wish to, we cannot escape its impact on our daily lives, nor the moral, social and economic dilemmas with which it confronts us. We may curse it or bless it, but we cannot ignore it.

Least of all can economists afford to ignore innovation, an essential condition of economic progress and a critical element in the competitive struggle of enterprises and of nation-states. In rejecting modern technology, Simone de Beauvoir is consistent in her deliberate preference for poverty. But most economists have tended to accept with Marshall that poverty is one of the principal causes of the degradation of a large part of mankind. Their preoccupation with problems of economic growth arose from the belief that the mass poverty of Asia, Africa and Latin America and the less severe poverty remaining in Europe and North America, was a preventable evil which could and should be diminished, and perhaps eventually eliminated.

Recently both the desirability and the feasibility of such an objective have been increasingly questioned. However, innovation is of importance not only for increasing the wealth of nations in the narrow sense of increased prosperity, but also in the more fundamental sense of enabling men to do things which have never been done before at all. It enables the whole quality of life to be changed for better or for worse. It can mean not merely more of the same goods but a pattern of goods and services which has not previously existed, except in the imagination.

Innovation is critical, therefore, not only for those who wish to accelerate or sustain the rate of economic growth in this and other countries, but also for those who are appalled by narrow preoccupation with the quantity of goods and wish to change the direction of economic advance, or concentrate on improving the quality of life. It is critical for the long-term conservation of resources and improvement of the environment. The prevention of most forms of pollution and the economic re-cycling of waste products are alike dependent on technological advance.

In the most general sense economists have always recognized the central importance of technological innovation for economic progress. The famous first chapter of Adam Smith's *Wealth of Nations* plunges immediately into discussion of 'improvements in machinery' and the way in which division of labour promotes specialized inventions. Marx's model of the capitalist economy ascribes a central role to technical innovation in capital goods—'the bourgeoisie cannot exist without constantly revolutionizing the means of production'. Marshall had no hesitation in describing 'knowledge' as the chief engine of progress in the economy. A standard pre-war text book states in the chapter on 'economic progress' that 'Our brief survey of economic expansion during the last 150 years or so seem to show that the main force was the progress of technique' (Benham, 1938, p. 319). Samuelson today comes to much the same conclusion.

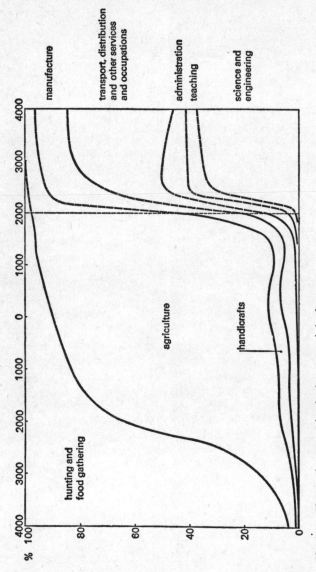

Fig. 1.1   Changes in occupation in the past and the future
Source: Bernal (1958)

it as primary, although operating in close association with other factors. Education and training of the labour force, efficient communications, additional capital investment, economies of scale, structural changes, plant reorganization, and the application of management skills may all be regarded as the systematic exploitation and 'follow-through' of scientific discovery and technological innovation. In the most fundamental sense the winning of new knowledge is the basis of human civilization.

Consequently there is ample justification for concentrating attention on the flow of new scientific ideas, inventions and innovations. Efforts to generate discoveries and inventions have been increasingly centred in specialized institutions in both 'planned' and 'market' economies—the 'Research and Experimental Development' network. This professionalized system is generally known by the abbreviated initials 'R & D' in Western countries and as 'NIOKR' in the USSR. Its growth is perhaps the most important social and economic change in twentieth-century industry. This book is primarily concerned with the innovations arising from the professional R & D system, and with the allocation of resources to this system. Its interaction with other 'knowledge' industries and with industrial production and marketing are of critical importance for any economy, but it is only recently that it has become the subject of systematic study. The policy adopted for R & D in any country, whether it is *implicit* in the sense of *'laissez-faire, laissez-innover'*, or *explicit* in the sense of national goals and strategies, constitutes the main element of policy for science and technology, or, more briefly, national science policy. A wider spectrum of scientific and technological services (STS) link the R & D system with production and routine technical activities. STS includes such activities as quality control, information services, survey and feasibility studies.[2] They are also essential for efficient innovation, and may predominate in the diffusion of technical change in many branches of industry.

Although government and university laboratories had existed earlier, it was only in the 1870s that the first specialized R & D laboratories were established in industry. The professional R & D system was barely recognized at all by economists in the nineteenth century and even in the early part of this century the young Schumpeter (1912), who gave innovation pride of place in his theory of economic development, treated the source of inventions as exogenous to the economy. We owe to Schumpeter the extremely important distinction between inventions and innovations, which has since been generally incorporated into economic theory. An *invention* is an idea, a sketch or model for a new or improved device, product, process or system. Such inventions may often (not always) be patented but they do not necessarily lead to technical *innovations*. In fact the majority do not. An *innovation* in the economic sense is accomplished only with the first *commerical* transaction involving the new product, process, system or device, although the word is used also to describe the whole process. The chain of events from invention or specification to social application is often long and hazardous. Schumpeter (1912, 1928 and 1942) always stressed the crucial role of the entrepreneur in this complex innovative process. But as Almarin Phillips (1971) has pointed out, it was only in his later work that he recognized the 'internalization' of much scientific and inventive activity within the firm. In his 1928 article he pointed out that the 'bureaucratic' management of innovation was replacing individualistic flair and that the large corporation was becoming the main vehicle for technical innovation in

---

[2] These activities are defined and described in UNESCO (1969).

the economy. This shift of emphasis from the early Schumpeter (Model 1) to the late Schumpeter (Model 2) will be discussed further in chapters 6 and 10. It reflected the real change which had taken place in the American economy between the two World Wars and the rapid growth of industrial R & D in large corporations during that period.

By the time of the outbreak of the Second World War there was already in existence an extensive network of organized research laboratories and related institutions in government, in universities and in industry, employing a full-time professional staff. This 'R & D industry' can be subjected to economic analysis like any other. Its 'output' is a flow of new information, both of a general character (the result of 'fundamental' or 'basic' research) and relating to specific applications ('applied' research). It is also a flow of models, sketches, designs, manuals and prototypes for new products, or of pilot plants and experimental rigs for new processes ('experimental development'). The inputs and outputs of this system are summarized in Table 1.1. But, of course, long before the twentieth century, experimental development work on new or improved products and processes was carried out in ordinary work-shops. When Boulton brought Watt's steam engine from the stage of laboratory invention to commercial production model, he most certainly carried out extensive 'research and development' at his Soho works, even if there was no department with that name.

The classical economists were well aware of the critical role of R & D in economic progress even though they used a different terminology. Adam Smith (1776) observed that improvements in machinery came both from the manufacturers of machines and from 'philosophers or men of speculation, whose trade is not to do anything but to observe everything'.

But although he had already noted the importance of 'natural philosophers' (the expression 'scientist' only came into use in the nineteenth century), in his day the advance of technology was largely due to the inventiveness of people working directly in the production process or immediately associated with it: '. . . a great part of the machines made use of in those manufactures in which labour is most subdivided, were originally the inventions of common workmen' (Smith, 1776, p. 8). Technical progress was rapid but the techniques were such that experience and mechanical ingenuity enabled many improvements to be made as a result of direct observation and small-scale experiment. Most of the patents in this period were taken out by 'mechanics' or 'engineers', who did their own 'development' work alongside production or privately.

The professionalization of industrial R & D

What is distinctive about modern industrial R & D is its scale, its scientific content and the extent of its professional specialization. A much greater part of technological progress is now attributable to research and development work performed in specialized laboratories or pilot plants by full-time qualified staff. It is this work which is recorded in R & D statistics. It was not practicable to measure the part-time and amateur inventive work of the nineteenth century. Thus our R & D statistics are really a measure of professionalization of this activity. This professionalization is associated with three main changes:

1 The increasingly scientific character of technology. This applies not only to chemical and electronic processes but often to mechanical processes as well. The

Japanese, who are probably the most advanced nation in the applications of electronics to mechanical engineering, have coined the word 'mechatronics', which aptly expresses this transition. A formal body of 'book learning' is usually necessary now for those who wish to advance the state of the art, as well as practical experience.

2 The growing complexity of technology, for example, the partial replacement of 'batch' and 'one-off' systems of production by 'flow' and 'mass' production lines. It is expensive and sometimes almost impossible to use the normal production line for experiments in large-scale plants. The physical separation of experimental development work into specialized institutions was often a necessity in such cases. The sheer number of components in some processes and products has similar effects in prototype and pilot plant work.

3 The general trend towards division of labour, noted by Adam Smith, which gave some advantages to the specialized research laboratories, with their own highly trained manpower, information services and scientific apparatus.

Starting in the chemical and electrical industries, these laboratories have become increasingly characteristic institutions. Like all changes in the division of labour, the specialization of the R & D function and other STS has given rise to serious social problems, as well as to the benefits, which Adam Smith observed. As we shall see, the departmental separation of R & D from the production line and from the marketing function in the firm gives rise to major management co-ordination problems. The rise of a professional 'R & D establishment' as a distinct social group may also lead to even more serious divisions and tensions in society, between those who generate new knowledge and others who may not understand it or may not want to see it applied. The R & D 'establishment' itself becomes a vested interest and political lobby, both in the industrial and in the military field. Some of these problems are discussed in the final section of this book.

The extent of specialization should not be exaggerated. Important inventions are still made by production engineers or private inventors, and with every new process many improvements are made by those who actually operate it. In some firms there are 'Technical' or 'Engineering' departments or 'OR' sections, whose function is often intermediate between R & D and Production and who may often contribute far more to the technical improvement of an existing process than the formal R & D department, more narrowly defined. But the balance has undoubtedly changed, and it is this specialization of the R & D function which justifies some such expression as the 'research revolution' to describe what has been happening in twentieth-century industry. During this time most large firms in the industrialized countries have set up their own full-time specialized R & D sections or departments. Until the late 1960s, R & D activities were expanding very rapidly in many countries, but during the 1970s growth slowed down somewhat, especially in the United Kingdom and the United States. However, even in the United States, *industrial* R & D continued to expand, and in Japan and Germany this expansion was still substantial. By 1980, there were well over half a million scientists and engineers working in all types of R & D in the United States (Table 1.2) and about two-thirds of these were working in industry. There were over 300,000 engineers and scientists working in the Japanese R & D system and over 100,000 each in Britain and the German Federal Republic (Table 1.2). Research and Development is an extremely skill-intensive activity, so that the numbers of supporting staff (clerical, technical etc.) are relatively small, typically, only one or two per scientist or

Table 1.2a   Resources devoted to R & D and size of GDP and labour force by country; 1979

	Expenditures		Personnel	
	GERD (NSE + SSH)	GDP^l	RSE^f	Total labour force
	US$ million	US$ billion	thousand FTE	thousands
United States	56,163	2,334.5	628.6	105,032
Japan	20,063	983.9	341.5^g	55,960
Germany	17,366	763.9	111.0^c	$22,697^c$
France	10,366	571.3	68.0^c	$26,074^c$
United Kingdom[e]	6,994	318.2	104.4	26,316
Italy	$2,669^i$	323.6	39.7^c	$22,034^c$
Netherlands[d]	2,563	130.2	17.4^{cg}	$4,880^c$
Switzerland	2,329	95.0	11.8^c	$2,935^c$
Canada	$2,134^h$	227.1	24.6^c	$10,578^c$
Sweden	$2,012^h$	106.4	14.8^h	4,268
Belgium	$1,516^i$	108.3	9.2^c	4,056
Australia[b]	920	95.4	22.5	6,313
Denmark	640	66.2	6.0	2,668
Norway	632	46.7	7.1	1,909
Austria[c]	589^j	47.9	..	3,038
Yugoslavia	586^k	68.1^k	22.4	..
Finland	449	41.4	7.4	2,308
Spain[a]	375	108.1	6.2^m	13,675
New Zealand[c]	322	31.5	—	1,230
Ireland	134	14.8	2.8	1,235
Greece	77	38.6	2.6^n	3,375
Portugal[d]	57	17.8	2.1	4,177
Iceland[c]	13	2.0	0.1	98

Note: Countries are ranked in decreasing order of gross intramural R & D expenditure (GERD) at current exchange rates which do not necessarily reflect the balance of R & D costs between countries.

[a] 1976.
[b] 1976/77.
[c] 1977.
[d] 1978.
[e] 1978/79.
[f] Or university graduate.
[g] Not in FTE.
[h] NSE only.

[i] Provisional.
[j] National data including funds to abroad. Business enterprise data include 'Lagerstättenforschung' but exclude co-operative research.
[k] At OECD exchange rate of 18.9961 Dinars = US $1.
[l] Weighted for fiscal years.
[m] Business enterprise and government sectors only.
[n] OECD estimate.

GERD: Gross expenditure on Research and Development
RSE: Research Scientists and Engineers
FTE: Full-time Equivalent
GDP: Gross Domestic Product
NSE: Natural Sciences and Engineering
SSH: Social Sciences and Humanities

Source: Non-R & D data provided by the Economic Statistics and National Accounts Division, OECD and *Labour Force Statistics 1968-1979*, OECD.
R & D Data: *Science and Technology Indicators* (1981), OECD.

Table 1.2b Total national R & D resources in the late 1970s (major countries)

		United States	Japan	Germany	United Kingdom	France
Total RSE (FTE)	thousands 1978	596.8	341.5[c]	111.0[a]	107.7[d]	68.0[a]
Of which working in industry	thousands 1978	412.4	157.3	66.2	68.1	31.8
Total R & D employment (FTE)	thousands 1978	[1,300.0][e]	575.1[c]	319.3	[300.0][d]	222.1[a]
GERD	billion PPP$ 1978	49.7	14.4	9.7	7.0	6.3
	billion current $ 1978	49.7	20.1	14.4	7.0	8.3
RSE/labour force	1977 per thousand	5.8	6.1	4.2	3.8[b]	3.0
R & D employment/ labour force	1977 per thousand	[12.7][e]	10.4	12.2	11.4[b]	9.8
GERD/GDP	% 1979	2.39	2.04	2.27	2.20[b]	1.81
Resources per RSE and per R & D worker						
Expenditure per RSE	thousand PPP$ 1978	83	42	80[a]	65	90[d]
Expenditure per R & D employee	thousand PPP$ 1978	[38][e]	25	28[d]	23	28[d]
Support ratio per RSE		[11][e]	7	19[a]	19	23[a]
R & D expenditure per capita population	PPP$	227	125	158	125	130

a 1977.
b 1978.
c Working mainly on R & D.

d Partially OECD estimate.
e OECD estimate-based on regression.
f Partially OECD estimate.

[ ] Very approximate.
ppp: purchasing power parity
Source: OECD

Table 1.2c Average annual growth rates in total resources devoted to R & D during the 1970s

	United States	Japan	Germany	United Kingdom	France
GERD, 1970–9	1.5	6.9	4.1[b]	1.9[c]	3.1[b]
RSE, 1970–9	2.2[a]	9.5	4.3[b]	2.1[d]	2.2[b]
Total R & D employment, 1970–9	..	2.2	2.5[b]	..	1.3[b]

[a] 1971–9.　　[b] 1969–77.　　[c] 1969–78.　　[d] 1971–8.

Source: OECD

engineer. Nevertheless, this means that there were over a million people working on R & D in Western Europe in 1980 and probably over two million in the Soviet Union and other East European countries. This represents an order of magnitude increase over the scale of R & D in the 1930s—in real terms. When J. D. Bernal (1939) pioneered the measurement of R & D in his classic book, *The Social Function of Science*, he advocated just such an increase in Britain, although for rather different objectives.

For the economist, it is obviously desirable to examine the operations of this R & D system from the standpoint of its efficiency in employing scarce resources. How can the flow of new information, inventions and innovations be improved? Could the scientific and engineering manpower employed in an industrial laboratory or a government research station be more effectively deployed elsewhere? Could the information required be obtained free or at a lower cost from abroad? Are part-time or amateur inventors or scientists sometimes more productive than full-time professionals? What kind of economies of scale are there in research or in development? Can the gestation period for innovations be shortened? What kind of firms are most likely to innovate and under what market conditions? What type of incentives stimulate invention and innovation most effectively? How are innovations diffused through the economy? In what ways do universities contribute to industrial innovation and how could this contribution be improved? These are the kind of questions which economists ask about the R & D system. They should also ask some more fundamental questions about the relationship of innovations to wider human values. Are the main goals of science and technology the most desirable way of using these resources?

There is a considerable resistance to looking at invention and research in this way. One result has been that most studies of invention and innovation have been written by biographers who tended to concentrate on the personal peculiarities of famous inventors and innovators and memorable anecdotes of their exploits. A mythology has grown up, stressing mainly the random accidental factors in the inventive and innovative process. Sometimes these myths depart altogether from reality as in the case of Watt and the steam from the kettle; in other cases they simply exaggerate the role of chance events as in the case of penicillin.

The treatment of R & D as an exogenous and largely uncontrollable force, operating independently of any policy, has been promoted in the past by both economists and scientists, though for different reasons. In either case it encouraged the 'black box' and 'magic wand' approach to science and technology, which not only discouraged attempts to understand the social process of innovation, but even endangered the whole future relationship between science, technology and society. What is not understood may often be feared, or become the object of hostility. Even though it may have an elaborate theoretical justification, an 'implicit' *laissez-faire* policy for science and technology appears to many as the operation of blind chance, like its counterpart, 'implicit' economic policy of reliance on the private market forces to optimize allocation of resources.

Polanyi (1962) made an interesting analogy between a free market economy and the basic research system, arguing that in both cases decisions must be completely de-centralized to get optimal results. In the one case only firms have the necessary information on which to base good decisions and in the other case only scientists. Like most economists, Polanyi accepted the need for a central government subsidy to basic research because the private market would not finance such an uncertain long-term investment, but he maintained that the scientists should be completely free to pursue whatever projects they thought best. Friedmann took the argument one stage further in maintaining that there was no need for government to finance basic research at all. Like all such arguments, they can be carried to the point of absurdity by over-zealous logic. The market mechanism can be a useful technique for allocating resources in certain rather specific circumstances, but it has its limitations, so that the definition and implementation of social priorities for science and technology cannot be left simply to the free play of market forces (Nelson, 1959 and 1977). The political system is inevitably involved and the full implications of this situation are taken up in chapter 9.

Modern technology

The 'research revolution' was not just a question of change in scale, it also involved a fundamental change in the relationship between society and technology. The very use of the word 'technology' usually carries the implication of a change in the way in which we organize our knowledge about productive techniques. If by 'technology' we mean simply that body of knowledge which relates to the production or acquisition of food, clothing, shelter and other human needs, then of course all human societies have used technology. It is perhaps the main characteristic which distinguishes humanity from other forms of animal life. But until recently knowledge of these 'arts and crafts', as they used to be called, was largely based on skills of hand and eye, and on practical experience which was transmitted from generation to generation by some sort of apprenticeship of 'learning by doing'.

The expression 'technology', with its connotation of a more formal and systematic body of learning, only came into general use when the techniques of production reached a stage of complexity where these traditional methods no longer sufficed. The older arts and crafts (or more primitive technologies) continue to exist side by side with the new 'technology', and it would be ridiculous to suggest that modern industry is now entirely a matter of science rather than craft. The 'heating and ventilating engineer' may still be a plumber, the 'tribologist' may still be a greaser and the 'food technologist' has not yet superseded the cook. They may never do so.

Nevertheless, there has been an extremely important change in the way in which we order our knowledge of the techniques used in producing, distributing and transporting goods. Some people call this change simply 'technology'; others prefer to talk about 'advanced technology', or 'high technology', to distinguish those branches of industry which depend on more formal scientific techniques than the older crafts. Because in a sense human societies have always had 'technology', some people see little new in modern technology. It will be argued here that this is a profound mistake and that the newer technologies are revolutionizing the relationships between science and society.

Some historians have argued that 'science' and 'technology' are two sub-systems which developed autonomously and with a considerable degree of independence from each other. Derek Price (1965) maintains that the two bodies of knowledge were generated by distinct professions in quite different ways and with largely independent traditions. The scientific community was concerned with discovery and with the publication of new knowledge in a form which would meet the professional criteria of their fellow scientists. Application was of secondary importance or not even considered. For the engineer or technologist on the other hand, publication was of secondary or negligible importance. His first concern was with the practical application and the professional recognition which came from the demonstration of a working device or design. Derek Price does not of course deny that 'science' and 'technology' have interacted very powerfully. He uses the simile of two dancing partners who each have their own steps although dancing to the same music. The development of the steam engine obviously influenced thermodynamics (to put it mildly), whilst scientific knowledge of electricity and magnetism was the basis for the electrical engineering industry. But each partner in the dance has his or her own interpretation and moves in a different way.

This simile can be a useful one, but if it is used to argue that nothing has changed since the nineteenth century in the relationship between science and technology, then it can be dangerously misleading. At the very least there are some new 'dances' and some of them are 'cheek to cheek'. The relationship has become very much more intimate, and the professional industrial R & D department is both cause and consequence of this new intimacy. Two very important empirical studies, one British (Gibbons and Johnston, 1972) and the other American (National Science Foundation, 1973) have demonstrated in some depth the importance of 'science' and communication with the scientific community for contemporary technical innovation. Since the relationship is one of interaction, the expression *'science-related'* technology is usually preferable to the expression *'science-based'* technology with its implication of an over-simplified one-way movement of ideas. Marx spoke of the machine as the 'point of entry' of science into the industrial system, but this expression might be used with far more justification about the R & D department.

Walsh *et al.* (1979), in their study of science and invention in the chemical industry, showed that there was a very close similarity in the patterns of growth of patenting activity by firms and the publication of scientific papers. Liebermann (1978) demonstrated that in the electronics industry scientists actually cited more recent papers from the fundamental physics journals than their colleagues in universities.

Other historians and economists, notably Hessen (1931), Musson and Robinson (1969) and Jewkes, Sawers and Stillerman (1958) have insisted that already in the seventeenth, eighteenth and nineteenth centuries, there was a great deal of interaction between science and industrial technology. There is much truth in this contention, but it does not alter the fact that professionalized R & D, carried out *within* industry itself, has put the relationship on a regular, systematic basis and on a far larger scale.

This change has affected especially the design of new products, but the new science-related technologies also affect the way in which improvements and changes are made in production. As has already been suggested, in the older industries these could be made predominantly 'at the bench' by direct participants in the production process. The subdivision of mechanical process did not remove this possibility. Indeed, as both Adam Smith and Marx noted, the workers themselves were often responsible for inventions leading to futher subdivision. But the introduction of flow processes in the chemical industry and of electronic control and automation in other branches of industry mean that improvements and changes now depend increasingly on an understanding of the process as a whole, which usually involves some grasp of theoretical scientific principles. It also means that experiments usually have to be made 'off-line' in a separate workshop or pilot plant, rather than 'on-line' by production engineers or operatives. 'Systems analysis' becomes important in its own right. All this has accentuated the relative importance of the specialized R & D group or Engineering or Technical Service department and diminished the relative importance of the 'ingenious mechanic'. In the newer industries R & D personnel, as well as other technical departments and OR sections, often have to spend a good deal of time 'trouble-shooting', that is resolving difficulties which arise in the normal production process and are referred back to them for solutions. This is not strictly R & D but it illustrates the changed position of production staff. The use of R & D personnel to start and control new production lines in the semiconductor industry is another indication of this change, as is the trial operation of new instruments and machines first of all by R & D personnel.

This can also be seen from the patent statistics for the various branches of industry. In mechanical engineering, applications from private individuals are still important by comparison with corporate patents, but in electronics and chemicals they are very few. The overall share has been declining since 1900 (OECD 1982).

The increasingly scientific content of technology and the increased subdivision and specialization within science itself have led to major problems of communication between specialist and non-specialist. These have been accentuated by the divisions within the educational system between the different disciplines and between the arts and the sciences. For many people these tendencies, together with some of the unpleasant features of modern industrialization, have increased the sense of alienation from modern technology to the point where they question the desirability of any further innovation. They feel that the whole system is like an uncontrollable and unpredictable juggernaut which is sweeping human society along in its wake. Instead of technology serving human beings it sometimes seems to be the other way about. The constant reiteration of the stock reply, 'You can't stop technical progress anyway', serves to reinforce rather than to diminish these fears.

As a result, the social mechanisms by which we monitor and control the direction and pace of technical change are one of the most critical problems of contemporary politics. In the final chapter of this book it is argued that a more explicit policy for science and technical innovation is increasingly necessary. It is also argued that the market demand mechanism for innovation in consumer goods and services has serious deficiencies. But it is by no means easy to understand or to control this complex system and the high degree of autonomy which it enjoys is partly the result of this difficulty. Socialist societies have not been particularly successful either.

This is not to deny that a pure *'laissez-innover'* system is unacceptable. Nor is it to deny the paramount importance of human values in deciding whether to promote or to halt particular new technical developments. Technical innovation need not be a purely random or arbitrary process but control depends upon understanding, and an important part of this understanding relates to economic aspects of the process, such as costs, return on investment, market structure, rate of growth and distribution of possible benefits. We still know far too little about these economic aspects of innovation, but slowly we are beginning to build up a body of systematic observations and generalizations, together with explanatory hypotheses which are supported to a varying extent by the empirical data. No doubt some of these hypotheses will be wholly or partly refuted or modified by future observations and experiments. As our knowledge extends so does the possibility of using innovations more satisfactorily.

Structure of the book

This book reflects the relatively elementary state of our present knowledge. The generalizations are tentative because they have been insufficiently tested and corroborated by applied research. Although the book describes the results of some of the empirical studies by European and American economists, it also poses some of the principal unsolved problems, in the hope that this will help to stimulate new thinking and research. Finally, the last part of the book raises some of the difficult policy issues which arise from the analysis.

The choice of a historical method of approach in the first part of the book is deliberate. The abstract 'representative firm' is a fictional device which is of little value in understanding the role of industrial R & D. In order to make useful generalizations about R & D in relation to firm behaviour it is essential to place the growth of this phenomenon firmly in a historical context. Robinson Crusoe is of little help, and a pure hypothetico-deductive approach is impotent without a preliminary process of observation and description. This is the purpose of Part One. It is designed to illustrate the three basic aspects of the rise of the professionalized industrial R & D system discussed above—growing complexity of technology, increased scale of processes and specialization of scientific work. Such historical description is of course intended to lead to the generation and examination of hypotheses in a systematic manner.

The whole of Part Two is devoted to an examination of the empirical evidence which might be held to support or refute various contemporary theories of innovation, particularly in relation to firm behaviour. The evidence which is used includes both the historical material cited in Part One and additional studies which have a bearing on the problems. Thus the main concern of Part One is with description and historical context, Part Two with analysis and Part Three with policy. Some readers may wish to skip the historical detail contained in Part One, but they will find that Parts Two and Three constantly revert to cases cited in Part One for illustration and support, and that some of the analytical problems are raised in a preliminary way during the course of Part One, at the end of each chapter. Some of the difficult *measurement* problems are first discussed in chapter 3.

Part One (chapters 2–4) deals in a historical–descriptive manner with research, invention and innovation in chemical and oil process plant and nuclear energy (chapter 2), synthetic materials (chapter 3) and electronics (chapter 4). The selection is not random. It is based on research projects which were carried out at the National Institute of Economic and Social Research in London and the Science Policy Research Unit at the University of Sussex in the 1960s and 1970s.

Professor Barna, who initiated the NIESR studies, analysed and compared the rate of growth of a large number of products in world industry and world trade in the years since the Second World War. This analysis led to the identification of three main 'clusters' of fast growing new products—electronics, synthetic materials and related organic chemicals. They had a growth rate well above the average for the whole post-war period, generally exceeding 10 per cent per annum compound. This may be confirmed by direct comparison of industrial and trade statistics of any of the advanced industrialized countries. New products and processes are generally the fastest growing product groups or 'clusters', although their growth may have slowed down a little once they reach the stage of inclusion in official statistics.

These industries (electronics, instruments, chemical and oil process plant, plastics, nuclear) barely existed before the twentieth century and are characteristic of the new research-based industries. All of them are 'research-intensive'; that is, they have a high ratio of professional R & D manpower in relation to total employment, or of R & D expenditures to net output (Table 1.3 and Figure 1.2). If military aircraft are excluded, then they account for about half of all industrial R & D in most countries. Moreover, firms in these industries have tended to 'invade' and 'colonize' the more traditional industries. This can be seen very clearly in the impact of the chemical industry on textiles, food and building materials, of the electronics industry on precision engineering, machine tools and publishing, and of the chemical process industry on metals and food processing. Even in such a traditional industry as pottery, a recent study (Machlin, 1973) has shown the extent to which technical change has resulted from science-based innovation in the supply of materials. It is the contention of this book that the industries described in chapters 2–4 represent the most important trends of technical change, not that they are an average or a random sample.

Schumpeter (1939) suggested that long cycles of economic development were based on the introduction of major new technologies into the economic system and this theme is taken up in the final chapter of this book. The first such cycle, at the beginning of the industrial reovlution, was based on the steam engine and various textile innovations; the second, in the middle of the nineteenth century, was based on railways and steel, and the third on electricity and the internal combustion engine. It could be maintained that the fourth 'Kondratiev' long wave, from the 1930s to the 1980s, has been based on electronics, new sectors of the chemical industry, jet aircraft and nuclear power (see Freeman, Clark and Soete, 1982). Thus the coverage in Part One of this book represents most of the main growth sectors of this long wave.

The selection of these industries also provides a reasonable balance in the coverage of final *product* innovations, *process* innovations, *energy* innovations and innovations in *materials*. This breadth of coverage is desirable because all four categories are essential for economic progress.

It is not possible in Part One to describe any of the innovations fully, as each one would merit a book in its own right. Some of the books which have been written are cited in the references. The intention here is to select only some of the most important characteristics of the innovations for discussion, from the standpoint of the economist. The treatment of technical aspects of the innovations is minimal, and so is the treatment of the personal characteristics of the inventors and innovators. Attention is concentrated on such questions as cost, patents, size of firm, marketing and time lags. What kind of firms made the principal innovations?

Table 1.3a　United States resources for research and development by industry (US$ million, 1978 and numbers of scientists and engineers)

Industry	Total R & D US$ million	Total R & D % net sales	Company-financed R & D	Federal-funded R & D	Full-time equivalent R & D scientists and engineers (thousands)			1978 per 1,000 employment
					1958	1978	1981	
Food and kindred products	428	0.4	..	..	4.8	7.5	7.5	8
Textiles and apparel	86	0.4	..	..	0.8	1.8	1.7	3
Lumber, wood products	136	0.7	136	Nil	0.8	2.2	2.3	6
Paper and allied products	394	1.0	..	..	1.7	7.3	8.2	13
Chemicals and allied products	3,598	3.6	3,232	365	31.0	49.4	52.8	42
Industrial	1,835	3.5	1,488	347	18.8	21.8	22.1	39
Drugs and medicines	1,282	6.3	..	..	5.1	19.9	21.1	63
Other chemicals	481	1.8	..	..	7.1	7.6	9.6	25
Petroleum refining	1,071	0.8	952	119	7.4	10.8	13.4	20
Rubber products	485	1.9	..	..	4.7	..	..	18
Stone, clay and glass	330	1.3	..	..	..	5.3	..	13
Primary metals	546	0.6	518	28	5.2	8.2	8.2	13
Ferrous	263	0.5	258	5	3.0	3.7	3.6	7
Non-ferrous	283	0.9	260	23	2.2	4.5	4.6	13

Fabricated metal products	397	1.1	360	37	8.3	7.5	9.3	11
Machinery	4,469	5.0	3,875	594	27.4	62.9	65.3	38
Office, computing etc.	3,129	11.7	2,577	552	..	42.9	46.1	76
Other machinery	1,340	2.1	1,300	42	..	20.0	19.2	38
Electrical equipment	6,739	6.2	3,769	2,970	47.9	89.2	105.8	40
Radio and TV	54	1.1	54	Nil	..	1.0	1.0	14
Electronic components	838	6.6	..	..	}22.3	..	..	51
Communication equipment	3,251	7.7	1,887	1,364		43.2	51.2	46
Other electrical	2,596	5.3	..	..	25.6	30.2	33.7	32
Motor vehicles	3,782	3.3	3,333	449	}15.0	33.5	30.4	24
Other transport equipment	131	1.4	..	..		2.0	1.2	14
Aircraft and missiles	7,700	12.3	1,863	5,837	58.7	88.6	96.1	87
Professional and scientific instruments	1,723	6.1	1,529	194	10.2	24.2	27.0	44
Scientific instruments	475	5.8	450	26	5.8	9.0	11.0	47
Optical, medical, photographic	1,248	6.2	1,080	168	4.4	15.2	16.3	43
Other manufacturing	280	0.7	272	9	}17.8	4.9	5.0	8
Non-manufacturing	1,094	..	594	500		16.4	20.9	17
TOTAL	33,400	3.1	22,098	11,302	234.8	427.8	469.9	28

Source: National Science Foundation.   .. not available

Table 1.3b Concentration of industrial R & D by sectors in major OECD countries: share of the engineering industries in total industrial R & D during the 1970s (% of total)

	United States		Japan		Germany		United Kingdom		France	
	1970	1979	1970	1979	1969	1977	1969	1978	1970	1978
Aerospace	28.9	22.3	..	..	7.0	7.2	23.1	18.3	20.5	17.1
Electrical	8.6	7.3	10.9	10.1	..	..	5.2	4.0	..	3.7
Electronics	14.8	12.8	13.9	13.2	..	..	15.5	20.1	..	20.4
Sub-total	23.4	20.1	24.8	23.3	28.1	26.2	20.6	24.1	21.2	24.1
Instruments	4.1	5.4	2.3	2.9	1.5	1.2	2.5	1.5	1.1	1.3
Machinery	9.6	5.1	8.8	7.0	} 7.6	} 12.5	7.1	4.7	5.9	3.6
Computers	..	9.5	2.8	2.8			3.4	6.2	5.4	4.7
Sub-total	13.7	18.9	13.9	12.6	9.1	14.1	13.0	12.3	12.3	9.6
Motor Vehicles	..	11.7	9.5	14.0	14.7	11.8	7.2	5.6	9.7	12.0
Shipbuilding	..	..	} 2.0	2.5	..	0.1	0.7	0.8	0.1	..
Other	..	0.4		0.2	..	0.2	0.0	..	0.1	0.4
Sub-total	8.8	12.1	11.5	16.7	14.8	12.2	7.9	6.4	9.9	12.4
Percentage of total BERD	74.8	73.4	50.2	52.6	59.0	59.7	64.6	61.1	63.9	63.2

BERD = Business Enterprise Research and Development   .. not available

Source: Science and Technology Indicators, OECD, 1981.

Table 1.3c Share of the chemical group in total industrial R & D

	United States		Japan		Germany		United Kingdom		France	
	1970	1979	1970	1979	1969	1977	1969	1978	1970	1978
Chemicals	7.1	6.8	15.8	11.7	{28.2	{26.4	9.9	9.8	9.7	9.9
Drugs	2.7	3.8	5.5	6.6			3.6	6.9	4.6	5.7
Petroleum products	2.9	3.2	1.2	1.0	0.5	0.5	1.7	1.3	3.7	3.1
Total	12.7	13.9	22.4	19.4	28.8	26.9	15.1	18.0	18.0	18.7

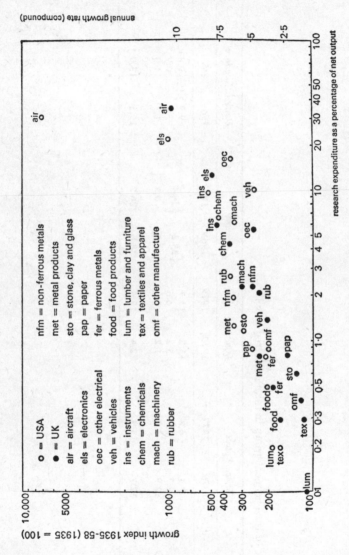

Fig. 1.2 Research expenditure as a percentage of net output in 1958, and growth of industries 1935–58

How far were they the result of professionalized R & D? How long did it take to develop and launch the new products and processes? How much did it cost? What were the expectations of management and the pressures which led to the decision to innovate? What are the implications for the theory of the firm?

Although the approach concentrates on the economic aspects, this does not mean that technical, psychological and other aspects of innovation are unimportant. Such an attitude would obviously be absurd. It would be a fair criticism that a more integrated theory of innovation is desirable, but it is beyond the scope of this short book. However, some of the wider social issues involved in policy for technical innovation are discussed in the concluding chapters.

The largely descriptive historical treatment of technical innovation in plastics, electronics, process plant and nuclear power in Part One is followed by an analytical treatment of some of the *general* implications for innovation theory in Parts Two and Three. Chapters 5–8 are concerned with problems of the *firm* in relation to innovation. Chapters 9 and 10 discuss some of the implications for *national* policies for science and technology.

In the analytical section it is argued that the professionalization of R & D described in Part One had far-reaching consequences on the nature of the competitive struggle between firms, both on the national and the world market. The factors which lead to success or failure in this new type of competitive struggle are discussed in chapter 5, and the implications for size of firm in chapter 6. In general the growth of industrial R & D has favoured the large firm and has contributed to the process of industrial concentration, but small new firms retain an advantage in some types of innovation. The giant international corporation has the great advantage of being able to spread the very high development costs of some kinds of innovation and the associated technical services over a very large sales volume. This is an enormous asset in industries such as computers, telecommunications, turbine generators, refineries, aircraft and drugs. But a high degree of uncertainty remains characteristic of technical innovation whether in large or small firms. The problems for the firm in coping with this high degree of uncertainty in managing innovation are discussed in chapter 7.

The type of groping and experimental decision-making characteristic of the innovation process is not compatible with theories of the firm which postulate a high degree of accuracy in investment calculations or extensive fore-knowledge of the consequences of the firm's behaviour. The uncertainty associated with innovation is such that differences of opinion about the desirability of alternative projects and strategies are the norm rather than the exception. This means that the firm is typically the arena of political debate between the advocates of alternative courses of action, and that power struggles will take place around these issues.

This leads to some reconsideration of the theory of the firm in chapter 8. The firm attempts to use R & D and other scientific and technical services to reduce the uncertainty which confronts it. But the nature of R & D is such that technical and market uncertainties remain despite its best efforts. Some types of R & D may indeed increase the uncertainty. Consequently, a high degree of instability will remain and decision-making in the firm will continue to resemble a process of 'muddling through' rather than the ordered, rational calculation beloved of neo-classical theory.

Such a conclusion has major implications for national policy. Not only have governments become heavily involved in assisting firms with their innovations, whether military or civil, but they are likely to have increasing responsibilities for 'technology assessment', that is for comprehensive social cost-benefit analysis of the probable consequences of technical change. The socialization of some of the risks and uncertainties of technical innovation is difficult to avoid because of the pressures of world competition, externalities and scale factors in R & D, and the adverse consequences of *'laissez-innover'*. Such socialization, however, carries with it the responsibility for the development of an *explicit* rather than an implicit national policy for science and technical innovation. Some problems associated with this major government responsibility are discussed in chapter 9.

It is argued there that in the US, the USSR, France and Britain the priorities of the 1945–80 period were largely determined by the Cold War. Government support for aircraft, nuclear and electronics R & D was both massive and effective. Firms in these industries became part of a special military–industrial complex, in which state-supported innovation was normal. Quite different priorities should be established in the last part of the twentieth century and national policy should be concerned to promote other kinds of innovation. A great deal of R & D will be needed to cope with environmental problems, to secure long-term supplies of cheap energy, to deal with natural resource limitations, to promote full employment, to develop much better transport and construction systems and generally to improve the quality of life in industrialized countries. Even more critical is R & D to deal with problems of underdevelopment. This redeployment of scarce R & D resources to meet the most urgent priorities is unlikely to occur solely as the result of short-term market factors. It must therefore be the main concern of national policy for science and technology, and increasingly of international policy.

Although reallocation of resources for R & D is essential, this is only a subsidiary aspect of the really important issue which is policy for technical *innovation*. Moreover, the analysis in Part Two concludes that the critical element in successful innovation is the social coupling mechanism, which links the professional R & D groups with the potential users of the innovations. The analysis suggests that the coupling mechanism which has been very effective for capital goods and for process and material innovations has been far less satisfactory for innovations in consumer goods and services. Finally, therefore, brief consideration is given to potential improvements in social mechanisms in these areas, which might contribute to a more genuine 'consumer sovereignty', and greater human satisfaction with the results of technical change.

2 Process Innovations

For over a century the chemical industry has enjoyed a high rate of productivity advance. The most important general change in the techniques of the industry has been the move from batch to flow processes of production. This has permitted very great economies of scale in plant construction and in labour costs for handling materials. Flow processes are also far more efficient than discontinuous batch processes in preventing heat losses, and in facilitating the monitoring and control of the chemical reactions. These advantages mean that unit costs of production have been drastically reduced for most of the major chemicals, and these reductions have affected not only labour costs but capital, energy and materials as well. At the same time, constant process improvements have led to higher quality and more uniform products. An example of the kind of economies which have been achieved is shown in Table 2.1, which compares production inputs per 100 gallons of petrol produced in the early US oil refineries before 1914 with the first fluid catalytic cracking processes in the 1940s and the improved versions available in the 1950s. The unit capacity of the early Burton process was about ninety barrels p.s.d.,[1] compared with about 13,000 for the early fluid installations and 36,000 for the later installations.

The most dramatic saving (of over 98 per cent) was in process labour costs, but almost equally impressive were the savings of over 80 per cent in capital costs and energy costs, and over 50 per cent in material inputs per unit of final output. It is this kind of technical progress which is fundamental to the growth of productivity and of the economy.

Table 2.1 Productivity comparison of the Burton and fluid catalytic cracking processes

Production inputs	Inputs per 100 gallons of gasoline produced		
	Burton process	Fluid process original installations	Fluid process later installations
Raw materials (gallons)	396.0	238.0	170.0
Capital ($, 1939 prices)	3.6	0.82	0.52
Process labour (man-hours)	1.61	0.09	0.02
Energy (millions of BTUs)	8.4	3.2	1.1

Source: Enos (1962a, p. 224).

The shift to flow production techniques was facilitated by six major developments during the nineteenth and twentieth centuries. These were:

1 The enormous growth of the market for the basic chemicals such as soda, ammonia, chlorine, sulphuric acid, ethylene and propylene. These 'building blocks' are used as intermediate materials for a great variety of other chemicals as well as in many other industrial applications, outside the chemical industry.

---

[1] Per stream-day; approximately 7.5 barrels per metric ton.

2 The switch in base materials for organic chemicals from coal derivatives to oil and natural gas. This stimulated the development of continuous processes and chemical complexes linked to refineries.
3 The increasing availability of electricity as a source of energy and the development of electrothermal and electrolytic processes. Faraday had demonstrated the electrolysis of salt in 1833 but it was not until the end of the century that cheap power became generally available and a large-scale process was developed.
4 Improvements in materials for plant construction and in components such as pumps, compressors, filters, valves and pressure vessels. These were essential to permit the use of large-scale processes, and more severe operating conditions such as high pressures and extremes of temperature.
5 The development of new instruments for monitoring and controlling flow processes, as well as for laboratory analysis and testing.
6 The application of basic scientific knowledge to the production processes, and the development of the new discipline of chemical engineering. The design of new flow processes was linked to physical chemistry whereas the old batch processes were often based on purely empirical knowledge and mechanical engineering.

All of these trends facilitated the growth of professional in house industrial R & D and were stimulated by it.

This chapter first outlines the characteristic chemical processes of the nineteenth century and the role of inventor–entrepreneurs in the development and innovation of these processes. This outline is necessarily very sketchy and intended only to give the background to the new pattern of process development which emerged in the German chemical industry in the latter part of the nineteenth century, and in the American oil refining industry in the early part of the twentieth century. The new industry of process design-engineering and contracting is then discussed including its relationship with the chemical and oil firms. The complexity of the design process for larger plants has now reached a point where hundreds of engineers and draughtsmen may be employed for months on one design.

This reaches its extreme limits in the design and development of nuclear reactors, discussed in the concluding section.

Nineteenth-century process innovation[2]

The characteristic plants of the early nineteenth century were the Leblanc soda and sulphuric acid plants, and the 'Alkali trades' was the name commonly given to the chemical industry as a whole. Leblanc was an ex-surgeon to the Duke of Orleans who developed his process on the basis of trial-and-error experiments on the Duke's estate from 1784–9, in response to the offer of a prize from the Academy of Sciences. Although he won the prize and was awarded a patent in 1791, he was an exceptionally unfortunate inventor–entrepreneur. His patron the Duke was guillotined and his St. Denis factory was expropriated. The Committee of Public Safety ordered the revocation of his patent and full publication of all the process know-how, and although his factory was returned to him in 1801, he committed suicide. However, his process dominated the European chemical industry for a hundred years.

[2] On the history of the chemical industry see Haber (1958, 1971); Hardie and Pratt (1966); Achilladelis (1973).

Some of the associated processes, such as sulphuric acid, were on a flow system but the kernel of the Leblanc system was batch production from coal furnaces, mixing one part of sodium sulphate, one part of calcium carbonate (chalk) with a half part of coal. This in turn was based on batch production of saltcake (sodium sulphate) by reacting sulphuric acid in a furnace with salt. Labour costs, material costs and fuel costs were all heavy, but big reductions in the price of alkali were achieved in the first half of the nineteenth century through improvements to the process.

The processes which replaced the Leblanc system and many of those which enlarged the whole scope of the chemical industry in the latter part of the nineteenth century were also mainly the work of inventor–entrepreneurs or individual inventors. Among the nineteenth-century inventor–entrepreneurs who established new firms to develop and exploit processes which they had invented or helped to invent were: the Swede Alfred Nobel (dynamite and other explosives), the French Brins brothers (industrial oxygen), the English chemists W. H. Perkin (aniline dyes) and John Bennett Lawes (superphosphates and other fertilizers), the German Linde (liquid air distillation), the Canadian T. L. Willson (electro-thermal production of calcium carbide), the Americans Castner and Down (industrial electrolysis), and the Belgian Ernest Solvay (ammonia-soda), as well as men like Parkes, Hyatt, Spitteler, Chardonnet and Baekeland who pioneered the early plastic materials, and to whose work reference will be made in the next chapter.

Almost all of these men were chemists, and although they often spent many years conducting research at their own expense and considerable personal risk (both physical[3] and financial), the scale of their experimental work was relatively small and the apparatus inexpensive. All of them attached great importance to patents (Nobel held 350 patents when he died), and most of them were successful in establishing a patent position so that production in other countries was either under a licensing arrangement or through affiliated companies. The successful enterprises which were built up became an important part of the twentieth-century chemical industry, whether they were absorbed into larger groupings (Nobel Divisions of ICI, IG Farben, British Oxygen Company), or remained independent companies (Dow, Solvay, Linde). The Nobel licensees in the US were Du Pont. The Solvay licensees in the UK were Brunner Mond and in the US the Solvay Process Corporation which subsequently merged into Allied Chemical and Dye.

Of the immediate successors to the Leblanc system, the Solvay process was the most important and may be regarded as the beginning of the modern heavy-chemical industry.

As is often the case when a product or process is threatened with extinction, many improvements were made[4] to the Leblanc process in the 1860s and 1870s and as late as 1887 the Chance process for the recovery of sulphur was highly successful. (It succeeded where Mond's process had failed largely because of improvements in carbonic acid pumps.)

Most of these process improvement inventions were made by British engineers and chemists who were involved in production problems in the Leblanc works in Cheshire and south Lancashire. Britain was the main centre of the Leblanc alkali industry from the 1830s to the 1880s, and had a thriving export business

[3] Nobel's brother was killed and explosions or fires affected Solvay seriously.
[4] A New Zealand scientist has christened this phenomenon the 'sailing ship effect' (Ward, 1967, p. 169).

to Germany and the United States. But the improvements were not enough to affect the basic technical superiority of the Solvay process and by 1890 most of the continental soda output was based on it and the whole of the infant American industry. Only in Britain did the old Leblanc process linger on, accounting for two-thirds of total production even in 1890, despite Mond's enterprise, as one of the first foreign licensees for the Solvay process.

The new pattern of process development in Germany[5]

The German and American chemical industries had no major investment in the Leblanc technology and did not suffer from problems of adjustment to the new techniques. The German industry in the 1870s had already established the new pattern of in-house R & D leading to the introduction of new products and processes. Bayer, Hoechst and BASF (Badische Anilin und Soda Fabrik) were among the first firms in the world to organize their own professional R & D laboratories. Although individual inventor–entrepreneurs made the major process innovations of the nineteenth century, by the end of the century the scale of flow process experimentation was putting it beyond the reach of the individual ingenious chemist, unless he enjoyed a large private fortune or the patronage of an established chemical firm.

The new pattern first became evident in the dyestuffs industry. Whereas some of the early synthetic dyestuffs were discovered and innovated by inventor–entrepreneurs in Britain, by the end of the century the lead had passed decisively to the German industry. Synthetic 'mauve' was discovered by the 19-year-old chemistry student W. H. Perkin in 1856 and within two years, with the help of his family, he had started a factory to make aniline dyes based on coal tar. This was followed by a succession of other aniline dyes, but with the alizarin dyes Perkin was beaten to the Patent Office by one day by BASF. However, he was still able to find an alternative patent-free route and launch manufacture of alizarin dyes in 1869. Both BASF and Hoechst had started with the manufacture of aniline dyes, and the principal distributor for BASF, Rudolph Knospe, had the exclusive sales rights for Perkins's aniline mauve from 1859.

In those days a number of outstanding German chemistry graduates were working in the British industry, but with the early successes of the infant German dye industry they began to return. Caro, who had been a successful inventor in Britain, became chief chemist at BASF in 1868 and it was he who launched alizarin dyestuffs manufacture just ahead of Perkin.

Two significant features of BASF policy from its foundation in 1864 were the insistence on integration of manufacture of intermediate and final chemicals and the concentration on creating a capability for plant and apparatus manufacture and repair. Not only did Caro play a leading part in establishing the Institute of German Chemists (VDC) but also of the Institute of German Engineers (VDI). Consequently, BASF was in a position to contribute to development of new chemical processes which required something more than individual brilliance, and depended on sustained co-operation between research scientists and qualified technologists. The establishment of a Department of Chemical Technology at Karlsruhe Technische Hochschule as well as the flow of graduate chemists from German universities very much facilitated their efforts to recruit highly qualified

[5] On the early history of BASF and Hoechst see BASF (1965), Baumler (1968), Beer (1959).

staff capable of product and process innovation. BASF, Hoechst and Bayer were managed by chemists who considered it their business to remain in close touch with the progress of university research. The work of Kekulé on the benzole molecule provided the theoretical basis for many of the major advances in coal-tar chemistry, which enabled the German dyestuffs industry to advance with extraordinary speed from 1870 to 1914. In 1880, Germany accounted for about one-third of world dyestuffs production, by 1900 about four-fifths, by which time there were 15,000 different patented dyestuff materials.

In the early period there was serious competition between the three leading German firms. All of them manufactured alizarin dyes as well as aniline dyes and all attempted the synthesis of indigo. But BASF helped to formulate the new Patent Law of 1877, which restricted price competition, and later informal understandings on markets led to the formal co-ordination of dyestuffs marketing through the 'IG' (Interessengemeinschaft) from 1904.

The synthesis of indigo is a good illustration of the new importance of systematic science-related process development. Bayer first took the lead in the 1870s in work on indigo synthesis, but without success. In 1880 Professor Baeyer, who had succeeded Liebig at Munich University, first produced synthetic indigo on a laboratory scale, followed by a whole series of indigo dyes. For this he was awarded a Nobel prize. Both Hoechst and BASF jointly took out patents on Baeyer's work, but the cost of production from his starting material far exceeded the price of the natural dye. BASF attempted to market the synthetic product but had to withdraw. In 1882 Professor Baeyer discovered a new synthesis, but again eight years' development work failed to yield an economic process. In 1890 Karl Heumann at Zurich Polytechnic discovered yet another synthesis and the patents were again acquired by BASF and Hoechst. Once more development work showed that the process, although technically feasible, was uneconomic.

After years of further effort, success was finally achieved at BASF, partly as a result of an accident during experiments on the oxidation of naphthalene, which at that time was extremely expensive. A thermometer accidentally broke and mercury flowed into the reactor vessel. It proved to be an ideal catalyst.

Hoechst had now fallen behind, but were able to retrieve their position as a result of collaboration with Degussa. At one time they had four different pilot plants each testing a different process. These trials showed that the Degussa process had decisive advantages and this became dominant at Hoechst in the early years of the twentieth century. The total cost of R & D on indigo synthesis from 1880 to 1897 was over 20 million marks. Once large-scale economic processes had been successfully developed at BASF and Hoechst, there was sharp price competition with the natural indigo dyes and Indian exports fell from 187,000 tons in 1895 to 11,000 tons in 1913. The price of the natural material fell from 11 marks per kilo to 6.50 marks over the same period.

By the end of the century, the German and Swiss chemical firms had established their supremacy as technical and market leaders, accounting for over 80 per cent of total world production. The leading Swiss firm, CIBA, maintained close research links with BASF and both they and the other Swiss leaders, Geigy and Sandoz, imported their basic chemicals and intermediates from Germany. The Swiss firms concentrated on research-based high quality dyes and drugs. By 1900 they were exporting 93 per cent of their output.

The new importance of an in-house process development capacity was evident not only in the commercial introduction of new products (whether plastics, dyestuffs or drugs), but also in the production of old-established basic chemicals. In the 1880s BASF developed the new 'contact' process for sulphuric acid which is still the basic technique for a large part of present-day production. The process was based on fundamental work by Wilhelm Ostwald at Leipzig in the field of physical chemistry, but depended on intensive industrial experiments with catalysts for the conversion to sulphur trioxide. BASF attempted to keep it secret, although in 1895 it was 'betrayed'—by an employee. When they took out patents in 1901 it was too late to prevent imitation, but the US General Chemical Company still found it worthwhile to take a licence from BASF to build the first contact plant in the US in 1906. As so often in the chemical industry process 'know-how' was important as well as the patents themselves.

But perhaps the most spectacular example of the successful marriage between fundamental chemistry and strong process-engineering capacity was the development of the Haber–Bosch process for synthetic nitrogenous fertilizers. Already, before 1900, BASF had been experimenting with various processes, but it was in 1908 that the forty-year-old Haber at Karlsruhe Technische Hochschule found a way to synthesize ammonia, based on reacting nitrogen and hydrogen at very high pressure and temperature in the presence of a catalyst.

BASF made an agreement with Haber for the development of his process and a strong development group led by Carl Bosch succeeded in designing and constructing the necessary pressure vessels and compressors to launch commercial production in 1913. This also involved improving and cheapening the catalyst. Some idea of the magnitude of the achievement may be gained from the fact that it took Brunner Mond around seven years to imitate the process (from 1919 to 1926), after an inspection of BASF plant in Oppau in 1919, and after other unsuccessful attempts during the First World War. It also took other chemical firms in France and US about as long, but Fauser at Montecatini was able to produce an improved process with a much higher conversion rate by 1925.

The first plant had an annual capacity of only about 10,000 tons, but many more were built in Germany during the war. BASF not only designed and developed the process, they also established an agricultural experimental station at Limburgerhof in 1914. The work of this station, and the numerous advisory centres which BASF established, made it possible for Germany to survive the effect of the blockade in cutting off Chilean nitrates, and to introduce the synthetic product very rapidly in German agriculture.

The success of the Haber–Bosch process had other important consequences. It gave BASF and the German chemical industry in general a lead in the development and operation of high pressure and catalytic processes. This proved extremely important not only for methanol (BASF process developed by Matthias Pier in 1922), but even more in the production of oil from brown coal, and in other high pressure processes developed in the 1930s.

The hydrogenation processes were sold to the US oil industry in the largest single lump-sum transaction ever known for process know-how in 1929, and led to the involvement of IG in major processes for the oil industry. Since developments in the oil-refining and chemical industry processes are so closely intertwined, we now turn to consideration of some refinery process innovations.

Process innovations in oil refining

Thermal cracking

The transactions between IG Farben and Standard Oil in the 1920s and 1930s were the culmination of an evolutionary development which started in 1900–13, with the first successful process for 'cracking' the heavy fractions of fuel oil to produce petrol (gasoline). It was in 1855 that a Yale chemistry professor had first demonstrated the phenomenon of cracking, but it was not until the twentieth century that this discovery found commercial application in a succession of revolutionary new processes.

These innovations have been very carefully documented by Enos (1962a). His book is one of the best which has been written on the history of innovation and this account draws heavily on his findings (see also Enos, 1962b). The background to the new processes was the rapid increase in demand for one of the light volatile products of the oil industry (petrol), a drastic fall in demand for another light product (kerosene), and a relative decline in demand for the heavier products (fuel oil). This was, of course, in turn linked to the growth of the automobile industry, and the replacement of the paraffin (kerosene) lamp by electricity. The characteristic pattern of demand for refinery products in the United States was rather different from the European pattern. In Europe the demand for the lighter products was and remains a smaller fraction of total demand, although it shifted in the same direction with a time lag.

The demand for an improved yield of refinery products was particularly strong in the Middle West, where the output of local oilfields was already declining by 1900, transport costs were high and competition from cheap coal was strong for the heavy residual fuel oil. It was generally known in the industry that intensive heating of the residual led to 'cracking' and a crude distillation process called 'coking' was in use. This amounted to cooking heavy fractions in an open vessel at atmospheric pressure, which gave a very low yield of lighter products.

The man who introduced the first really successful commercial cracking process was William Burton, who took a Ph.D in chemistry at Johns Hopkins University in 1889. In his boyhood he already had his own chemical laboratory and was befriended by Brush, the electrical inventor. Perhaps the most significant fact about his first industrial employment is that he was deliberately recruited and appointed by Standard Oil to run a laboratory at the Whiting refinery in the Indiana subsidiary. Although the laboratory was in an old farmhouse and he had to make many of his own instruments, he was able to make a number of improvements in the refinery methods. As a result, he was rapidly promoted to be superintendent of the refinery and two other Ph.Ds were appointed to the laboratory.

As refinery manager Burton was able to command the resources necessary for pilot plant experimental development in 1909 and 1910. He and his colleagues were able systematically to test out the results of cracking at various temperatures and pressures on different fractions of oil.

Their experiments at higher pressures were conducted at considerable personal risk, as the equipment available was primitive, and knowledge of high-pressure work only embryonic. They were limited by the size of plates and by the riveting techniques to a relatively small scale of operations and to what now appear as relatively low pressures. Nevertheless, they were successful in developing a much-improved process for gas-oil. But in 1910 the parent company refused to authorize expenditure of a million dollars to build the first plant, because of fears of explosions.

However, in 1911, as a result of an antitrust decree, Standard Oil of Indiana was divorced from the parent company and the new board authorized the expenditure. Production began in 1913 and the plant was highly successful, doubling the yield of petrol (gasoline). Marketing problems due to the colour and smell of the product had to be overcome, but Enos calculated that the total profits generated by the process for Indiana Standard were 123 million dollars from 1913 to 1922; the reduction in cost initially was 28 per cent, and ultimately with various improvements (mainly inventions patented by employees) about 50 per cent.

Enos estimates that the development cost of $236,000 was paid back ten times over in the first year of operation—1913. Subsequently, the royalty income alone amounted to over $20 million. The patent position, both on the original invention and the improvement inventions, was strong, and Indiana charged 25 per cent of the profits from using the process. Altogether nineteen companies were licensed by 1921, but they were restricted to selling in certain areas under the terms of the agreements. Moreover, they acquired the patent rights only, with no accompanying technical know-how.

These circumstances provided a significant stimulus to the development of alternative processes as well as leading to the collapse of many small refineries. In 1920 four and in 1921 five more new cracking processes were introduced, which gives some idea of the intensity of the inventive effort which followed the extremely profitable Burton process. Of these the most successful were the Dubbs process and the 'Tube and Tank' process. The Dubbs process is of particular interest because it led ultimately to the formation of a unique specialist process development company—the Universal Oil Products Company (UOP)—which has played a critical part in the subsequent history of the oil refining industry. The Cross process also led to the establishment of a process company but this did not survive.

Jesse Dubbs was manager of a small and independent Californian refinery and his son Carbon Petroleum Dubbs (believe it or not) had also managed a refinery. C. P. Dubbs was working for Standard Asphalt when this company was acquired by J. Ogden Armour, who wished to find new outlets for his enormous personal fortune from meat-packing. Under his influence Standard Asphalt sought and acquired all the patents held by Dubbs Senior. These included some which appeared to offer the possibility of a flow process for cracking and a way of bypassing the Burton patents. This was not the original intention of the inventor who was concerned mainly with the peculiar features of refining and marketing Californian crude oil. As a result of rather loose specifications, several inventions in 1909 could claim originality, and this led to many conflicts.

A new company (UOP) was set up to hold the patents and to develop processes. Dubbs had a 30 per cent holding and Armour 20 per cent. Dubbs Junior succeeded in developing and patenting a method of recirculating the residual heavy fraction back to the cracking coil. His collaboration in the UOP research laboratory with Dr Egloff,[7] an outstanding chemist, led to the successful construction of a pilot plant in 1918, which was displayed to Shell in 1919.

However, teething troubles and an explosion in 1921 on a plant being built for Shell necessitated considerable redesign and it was not until 1923 that the process became established. The total development costs had been about $6 million, which proved a heavy drain even for Armour, who went brankrupt in 1922.

---

[7] Ultimately Egloff became Director of Research and held 300 patents.

The new process proved superior to the Burton process in many different ways—most of them resulting from the inherent advantages of a flow process over a batch process. Unit size and capacity were vastly increased with a corresponding reduction in labour costs. Again, many cost-reducing improvements were introduced once the process was in operation. Electric welding permitted a big improvement in the size and performance of vessels and tubing. One of the most important developments was the introduction of a pump which could perform satisfactorily under the high temperatures and pressures of the recycle stream. The original patent for the hot-oil pump was taken out by Shell, but UOP's engineering department improved the design. This invention alone increased capacity by 40 per cent.

Shell were given a 25 per cent reduction in their royalty rate and this illustrates an important aspect of UOP policy. As a process-development company they insisted on reciprocal exchange of know-how and process improvements and they also provided technical assistance and performance guarantees. They insisted on making the process available to all comers, without discrimination, even though they had requests for exclusive licensing. Originally the Dubbs process was licensed to Shell and many small West Coast refiners, but in 1924 Standard Oil of California also took a licence. By 1930 both they and Shell were paying over $2 million per annum in royalties to UOP. They had not expected to pay on this scale, as a lawsuit initiated in 1914 by UOP against Indiana Standard was generally expected to lead to the nullification of both sets of patents. However, the legal process dragged on for fifteen years and was only finally resolved in 1931 when Shell, California, and Indiana with two other oil majors bought UOP for $25 million. The royalty rate was gradually reduced from the original flat rate of 51 cents per barrel to 10 cents in 1934, 5 cents in 1938 and 3 cents in 1944. By this time superior processes had been developed and in particular the royalty rate on the fluid catalytic-cracking process had been announced at 5 cents per barrel.

However, UOP continued to follow a relatively independent policy under the new regime. Hiram Halle, who had been an exceptionally able and enterprising President since 1916, was to stay in office for another fifteen years under the purchase agreement terms. He insisted on an offensive R & D policy and accepted Egloff's proposal to recruit some of the best scientists in the world to work on catalysis. Egloff argued that contemporary cracking processes were close to maximum efficiency and a new breakthrough would be needed. As a result UOP recruited Ipatieff, an outstanding Russian chemist, Tropsch and several other German chemists with experience of IG processes. On the basis of this far-sighted policy UOP were responsible for many other process improvements, and made a major contribution to the fluid catalytic cracking process. One of their many contributions which was adopted by almost all refineries was platinum catalytic reforming ('platforming'), introduced in 1949. The oil industry came to accept UOP's unique role and in 1944 transferred ownership to the American Chemical Society, but in 1958 it once more became an independent company, conducting R & D, selling process designs and providing technical consultancy—a firm producing and trading in new knowledge.

Whilst the Dubbs process was largely developed by an independent process company, the other major continuous flow process of the 1920s—the 'Tube and Tank'—was the result of a deliberate decision by the largest oil company (Standard Oil of New Jersey) to set up its own specialized R & D department. This was originally named the Standard Development Company, but from 1955 became

Esso (later Exxon) Research and Engineering. Established in 1919, by 1920 it already employed over fifty people and many hundreds by the 1930s. It employed university consultants and also attached great importance to patent work, because of the tangled patent situation with many overlapping patents in the 1920s. The inventor who played the biggest part in developing the 'Tube and Tank' process was Edgar Clark, who had been one of Burton's collaborators, but the process was deliberately *not* given the name of an individual as it was felt to be largely the result of team-work. Although the process was developed fairly quickly and in operation by 1921, its main advantages only became apparent later during the course of many improvements made by the development group. These made it possible to process a great variety of crude stocks, which could not be handled by the Burton process, and to raise the operating pressure from 95 to 100 lb per square inch by 1930. A big improvement in performance was made as a result of systematic analysis and re-design of heat exchange in the recycling system and the introduction of the Pacific hot oil centrifugal pump in 1929. In spite of heavy legal and patent costs (over $1 million) the process was extremely profitable, making about $300 million. The complicated patent litigation was resolved by the formation of a licence exchange agreement in 1923, and, as we have seen, the purchase of UOP by five oil companies in 1931.

Catalytic cracking

It was widely realized in the 1920s that further progress in cracking was likely to come from catalytic techniques. Already in 1915 the Gulf Refining Company had attempted to introduce catalytic cracking with an aluminium chloride catalyst, but the process was abandoned because of the high costs, and the lack of a satis-factory method for regenerating the catalyst which became clogged with an accumulation of carbon. Several oil companies continued to experiment with catalysis in the 1920s, but the first successful process was invented outside the industry by a wealthy French engineer, Eugène Houdry. His father was a manu-facturer of structural steel, and Houdry at first entered the family business, but he gave it up in 1923 to devote himself completely to experimental work on new types of motor fuel, originally from lignite. In 1925 he became interested in catalysis of oil fractions, and with the assistance of friends, who included both chemists and engineers, he experimented with hundreds of catalysts. By 1927 he had succeeded in cracking experiments with a type of clay (silicone and aluminium oxides).

Houdry published his findings and the major oil companies, including Standard Oil (New Jersey), Shell and Anglo-Iranian (BP) sent technical men to visit his laboratory. However, they were sceptical of the possibilities of introducing the process commercially, and in the case of Standard Oil had greater hopes of other process developments. Houdry continued to spend his private fortune, but reached a point where he could not make further progress without additional financial support, and even more without the participation of an oil company to help in the design, construction and test of the pilot plant equipment at a refinery. He was also disappointed over the withdrawal of official French Government support for pilot plant work on another lignite process in which he was interested. These circumstances led to his emigration to the United States and the formation there of the Houdry Process Corporation, jointly with the Socony Vacuum Oil Company in 1930. The development of the process continued to give difficulty and in 1932 he reached a new agreement with Sun Oil, who gave more substantial reasearch and engineering support, as well as taking a one-third holding in the Houdry Corporation. It was only after the expenditure of about $11 million that the process was successfully introduced by Socony Vacuum and Sun in 1936–7. Of this amount, $3 million came from Houdry's own private fortune and about $4 million each from Sun and Socony.

This outline indicates clearly the difficulties confronting the independent inventor in process development, even when he had a large private fortune and almost unlimited perserverance and enthusiasm. The scale and complexities of pilot plant work had reached a point in the 1920s where collaboration with an established oil or chemical company was often essential for experimental development work. A critical factor in the ultimate success of the Houdry process, as so often in process development, was the introduction of a new piece of engineering equipment—the turbo-compressor based on a Swiss design. This made the regeneration cycle for the catalyst economic for the first time, and was introduced as a result of experiments by Sun R & D staff, and a technical mission to Switzerland. The Houdry process was ultimately not only less costly than thermal cracking, it also produced much better quality products, and unexpectedly, very good aviation fuel.

The successful introduction of a catalytic cracking process by Sun Oil and Socony Vacuum was a powerful stimulus to the parallel work of Standard Oil of New Jersey. Licensing negotiations indicated that the Houdry Process Corporation expected to get about $50 million from Jersey if the Houdry process was adopted as their standard cracking process. This very tough attitude led Jersey to reject this possibility and concentrate instead on developing their own process. They were in any case inclined to do this because of the limitations of the fixed-bed Houdry process, their own R & D capability, and the know-how they had acquired from IG Farben.

Already in the First World War Bergius at BASF had developed a process for synthesis of oil from lignite (brown coal) by hydrogenation at high pressure. It seemed that this process might have even greater application for the heavy oil fractions than for lignite, and Standard Development reached a technical know-how exchange agreement with IG after a visit to the BASF plant in 1927. Standard agreed to supply IG with information on their heavy oil work, and in 1929, convinced of their future success, they paid $35 million for the world rights (except Germany) to the IG patents and processes. Standard had originally expected that the process would be widely applied for low-grade crude oil stocks, but although it was successful in the plants they built in 1930 and 1931, there were still ample world supplies of high grade crude oil. An indication of the strength of the German technical lead in high pressure process work at this time was the fact that all the valves and compressors had to be imported from Germany, because American manufacturers were not able to supply the equipment for work at pressures of 400 lb per square inch. However, what Standard did acquire was substantial experience of catalytic processes, and a very strong patent position. In the early 1930s they had already developed the 'Suspensoid' process for catalytic cracking, using a powdered catalyst, and had experimented with several other techniques.

Consequently, Standard were in a good position to challenge the Houdry process by 'leap-frogging' to a better one. From the outset it was recognized that a fully continuous flow process would be far better than the semi-continuous Houdry fixed-bed catalyst system. Another major aim was to establish a process which could be used for a wide variety of crude stocks and not limited to higher grades.

In pursuing these aims they made common cause with other oil companies and process companies who felt themselves threatened by the Houdry process. In 1938 a group was formed known as Catalytic Research Associates, consisting originally of Kellogg, IG Farben, Indiana Standard and Jersey Standard and soon joined by Shell, Anglo-Iranian (BP), Texaco and UOP. Interestingly, it was Kellogg, a process design and construction company, which took the initiative in convening the first meeting in London which led to the joint research programme. The group (without IG) commanded the resources of R & D facilities employing about 1000 people (400 in Standard Development) and the work demanded the co-operation of specialists in many different fields.

The collaborative R & D, which was carried out from 1938 to 1942 to develop the fluid catalytic-cracking process, was one of the largest single programmes before the atom bomb. Jersey Standard made the greatest contribution and this was recognized in the agreements which were ultimately made on patents and royalties when the process was successful in 1942 (Table 2.2). However, the two process companies, UOP and Kellogg, also made very important contributions. UOP had already developed its own catalytic processes when it joined the group and was able to contribute substantially to the patent pool. Perhaps the decisive contribution to the process was the development of the fluid bed of fine particles of catalyst propelled in the stream of oil vapours. This was achieved at Jersey Standard with the assistance of MIT Chemical Engineering Department (including two graduate theses).

Table 2.2   Patents and royalties in the fluid catalytic-cracking process

Company	Patents contributed	Approximate share of royalties from process (%)
Standard Oil, New Jersey	296	39
Standard Oil, Indiana	96	2
Shell	38	7
Texaco	55	2
UOP	239	32
Kellogg	57	17

Source: Enos (1962a), Table 4, p. 217.

Although the total costs of R & D were ultimately over $30 million, the process was extremely profitable. Jersey Standard had received over $30 million in royalties alone by 1956. By this time the process accounted for over half of total cracking capacity and had prevented the Houdry process from ever achieving more than a ten per cent share. The original fixed-bed process declined rapidly after 1943, but improved versions (TCC and Houdriflow) continued to compete. The fluid process was a triumph of the big battalions.

Scale of plant and the process plant contractor

In both oil refineries and in the chemical industry proper, new process development had become an extremely expensive business. All the main cracking processes cost more than $1 million to develop except the Burton process developed before the First World War. The two main catalytic processes cost more than $15 million

(Table 2.3). The high costs arose mainly from the expense of pilot plant work, and the complexity of flow processes. Specialist groups were needed to cope with the design and engineering problems arising in each part of the plant, as well as in the overall design and heat transfer problems. At the same time economies of scale and the enormous operating advantages of flow processes put a premium on scaling-up existing processes.

Table 2.3   Estimated expenditure of time and money in developing new cracking processes[a]

Process	Development of new process		Major improvements to new process		Total	
	Time	Estimated cost $000	Time	Estimated cost $000	Time	Estimated cost $000
Burton	1909–13	92	1914–17	144	1909–17	236
Dubbs	1917–22	6,000	1923–31	> 1,000	1909–31	> 7,000
'tube and tank'	1918–23	600[b]	1924–31	2,612	1913–31	3,487
Houdry	1925–36	11,000[c]	1937–42	n.a.	1923–42	> 11,000
fluid	1938–41	15,000	1942–52	> 15,000	1928–52	> 30,000
TCC and Houdriflow	1935–43	1,150	1944–50	3,850	1935–50	5,000

a Excluding preliminary activities (background research and patent acquisitions).
b Including $100,000 legal expenses.
c Including some lignite process costs.
Source: Enos (1962a), p. 238.

The first Dubbs process unit had a capacity of 500 barrels per day; by 1931 this had increased to 4,000 barrels per day. The biggest Houdry units had a capacity of 20,000 barrels per day; by 1956 fluid catalytic crackers had a capacity of 100,000 barrels per day. At the same time the optimal size of plant for many other chemicals was rapidly increasing, for example, ethylene and ammonia plant, which typically had a capacity of 30,000 tons per annum just after the war, were being installed with a capacity ten times as great by 1965, and with much lower costs of production (Table 2.4).

The technical economies of scale in large process plants arise to a considerable extent from reduced capital cost, due to the simple fact that capacity is a function of volume while capital cost is a function of surface area. Whilst the volume of a cylinder is $\pi r^2 h$, the surface area (of the metal required to construct it) is $2\pi r h$. This means that as the volume is increased the amount of metal increases by only about half as much. Most parts of a chemical plant are columns, cylinders, pipes, spheres, etc., so that the 'plant factor' is usually about six-tenths. Where repetition of batch processes is involved then the 'plant factor' will be much closer to one (i.e. no technical economies of scale). But as we have seen, large scale of operations not only yields great economies in capital costs, but may also yield savings in supervisory labour, maintenance labour, operating labour, energy, feedstock, etc. (Similar considerations apply to oil tankers.)

However, there were signs in the 1970s that 'diseconomies' of scale were beginning to affect the construction and operation of some of the largest plants. During a period of very rapid economic growth, such as the 1950s and 1960s, new and larger plants can quickly be brought into full-capacity working provided the technical

Table 2.4   Influence of capacity on production cost (in £ thousand) of ethylene, 1963

Production costs	Capacity and output, 000 tons		
	50	*100*	*300*
capital investment			
battery limits plant	2200	3100	5400
off-site facilities	650	900	1600
total (excluding working capital)	2850	4000	7000
current costs per year			
net feedstock	250	500	1500
chemicals	50	100	300
utilities	300	600	1800
operating labour and supervision	50	50	50
maintenance at 4 per cent of battery limits plant cost	90	125	215
overheads at 4 per cent of battery limits plant cost	90	125	215
depreciation	250	355	620
total	1080	1855	4700
current costs (£) per ton of ethylene	21.6	18.6	15.7

Source: Wynn and Rutherford (1964).

problems of larger scale design are satisfactorily resolved. This means that during such periods the operation of 'Verdoorn's Law' (which associates growth of productivity with growth of total output) can be explained in terms of technical progress associated with increasing scale of plant and similar scale economies. But the slow-down in economic growth in the 1970s, together with the bunching of major investment decisions affecting very large plants, led to the emergence of serious over-capacity in several branches of organic chemical production, and especially in synthetic fibres and plastics. Prolonged below-capacity working has an extremely adverse effect on the economics of large plant operation because of the very high fixed capital costs. Such below-capacity working may be induced either by a failure of the market to grow sufficiently rapidly or by technical problems in the construction, commissioning and operation of new large types of plant. In either case, losses can be on a very large scale. Similar problems of diseconomy of very large plants appear to have set in during the 1960s and 1970s in electric power generation after a period of rapid scaling-up in size of power stations. Nuclear power was especially badly hit by a combination of slower-than-expected market growth and serious technical and safety problems.

But despite these diseconomies of scale affecting several highly capital-intensive industries, the combined effect of the shift to flow process, the increasing scale of operations, the use of catalytic processes, the complexity of plant design and the use of petrochemicals as the basic material has been to give the large chemical and oil companies a predominant position in new process development since the First World War. This must, however, be qualified by the observation that a new group of companies has emerged, which is also making a significant contribution: the specialist design, development and construction company (plant contractors and process companies).

The NIESR survey of 6000 chemical plants erected in the 1960s (Freeman, Curnow, Fuller, Robertson and Whittaker, 1968) showed that while some of the largest chemical firms still preferred to design, engineer, procure and construct their own new plants with their own 'captive' design-engineering organizations, the majority of plants are now engineered and built by process plant contractors. These results were confirmed by two later surveys in the 1970s and 1980s. All of these surveys pointed to the fact that the design and construction of new process plants all over the world is now a huge industry carrying with it a large volume of associated exports of mechanical equipment and instruments. This world-wide business is now dominated by the specialized plant contractors, rather than the chemical firms. However, the role of the chemical firms is still a crucial one, since they originate most of the major technical innovations in process design (Freeman *et al.*, 1968 and Mansfield, 1977, chapter 3) which are subsequently exploited by the contractors under appropriate licensing and know-how arrangements. The oil industry in the United States (through firms like Kellogg, Foster–Wheeler and UOP) pioneered this change, and although chemical firms try to safeguard their process secrets by keeping contractors out of sensitive areas, they are increasingly using their services.

Only a few of the contractors have strong research and development of their own, and are capable of designing and developing their own new processes. The disparity in technical strength is clearly shown by the patent statistics (Table 2.5). Mostly they depend on licensing processes from the big oil and chemical companies. But as a result of their experience in detailed design, engineering and construction of many process plants in different parts of the world for a variety of clients, they are often able to suggest and implement minor process improvements, and sometimes major ones. Moreover, the NIESR plant index showed that although over 80 per cent of new plants used processes originally developed by the major chemical and oil companies, a significant minority were either originated by specialist process companies or embodied modifications to basic processes developed by contractors, and marketed as proprietary processes.

Excluding construction site labour, plant contractors typically employ over half their staff on process design and 'engineering' (i.e. detailed flow sheets and drawings) and the remainder on procurement, sales and administration, with very small numbers on research and development. But a few contractors, such as Scientific Design and Lurgi, and process companies, such as UOP and Houdry, have a significant R & D activity. Although their total size is relatively small compared with the chemical companies they are sometimes able to afford the expenditure required to develop a new process. Scientific Design developed new catalysts for producing maleic anhydride and ethylene oxide in the early 1950s, and as a result their processes accounted for over a third of world capacity by the mid-1960s. But this is exceptional, and in each case development costs were over $1 million. Much more typical is the situation where a process company or a contractor collaborates with a large chemical or oil company in process development or improvements. The examples of Houdry and UOP have already been discussed and American companies have generally been more ready for this kind of collaboration than European.

Mansfield (1977) has shown that in the United States the contribution of the four largest chemical firms to major innovations was greater for product innovations than for process innovations in the period from 1930 to 1971 (1966 for product innovations). But in any case these firms accounted for over 55 per cent of

Table 2.5   Patents taken out in London by chemical companies and by contractors, July 1959–November 1966[a]

United States	United Kingdom	West Germany	France	Italy
Part A: Chemical companies				
Du Pont 1731	ICI 2998	Bayer 2161	Rhône–Poulenc 307	Montecatini 709
Esso Research & Engineering 1191	BP 641	Farbwerke Hoechst 1431	St. Gobain 231	
Union Carbide 1136	Distillers 491	BASF 1136	Institut Français Petrole 92	
Monsanto 723	British Oxygen 379	Wackerchemie GmbH 195	Solvay 90	
Dow Chemical 632	Courtaulds 277	Chemische Werke Hüls 166	Soc. Nationale des Pétroles Acquitaine 22	
Rohm & Haas 399	Shell Research 224	Ruhrchemie AG 134		
Olin Mathieson Chemical 358	Albright & Wilson 100	Deutsche Erdöl AG 21		
Allied Chemical 350	Laporte 96	Scholvenchemie AG 18		
W. R. Grace 350	Fisons 87			
Hercules Powder 191				
Ethyl 145				
Part B: Contractors				
Foster–Wheeler 108	Simon–Carves 106	Lurgi (Metallgesell.) 246 (377)	Heurtey 22	SNAM Progetti 24
Scientific Design 85	APV 67	Linde 141 (236)	L'Air Liquide 21	Oronzio de Nora 21
M. W. Kellogg & Pullman (chemical patents only) 59	Power Gas 38	Klockner–Humboldt-Deutz 131 (42)	Krebs et Cie 2	Impianti Elettro-chemical 7
Air Products & Chemicals 47	Woodhall Duckham 35			
	Humphreys & Glasgow 25			
	Whessoe 21			
	Matthew Hall Eng. 17			

Chemico	46	Petrocarbon Devt	14	Heinrich Koppers	117 (65)
Lummus	26	George Wimpey	6	Didier Werke	33 (15)
Hydrocarbon Research	13	Constructors John Brown	3	Pintsch Bamag	29 (53)
Houdry Process	9			Otto	22 (42)
Fluor	8			F. Uhde	19 (29)
Badger	7			Chemiebau Dr A Zieren	15 (18)
Chemical & Ind.	6			Firma Carl Still	14 (42)
Ralph M. Parsons	6			Edeleanu	13 (18)
Stone & Webster Engineering	3			Hans J. Zimmer	(41)

a Patents taken out in London and therefore with a national bias to British-based firms. German patent statistics are shown for German contractors for comparative purposes. The German statistics *exclude* coke ovens and coal installations. Patent protection for major innovations is normally sought in all highly industrialized countries, but for minor inventions protection is usually sought in the first instance in the home country, and only later in other countries if the expense appears justified. The activity of subsidiary companies adds a further complication; overseas firms with United Kingdom subsidiaries would be more likely to take out London patents at an early stage.

Source: Name Index to Complete Specifications, Patent Office, London. Figures in parentheses from Patent Office, Munich.

all innovations, except in the case of process innovations from 1950 to 1971, when their contribution fell to just over 40 per cent. This fall can be attributed to the increased contribution of the process plant contractors, such as UOP and Scientific Design, and of the oil companies, such as Shell. The chemical industry in the United States is less concentrated than in Europe and Mansfield's study provides valuable independent confirmation of some of the main conclusions of the NIESR/SPRU work.

Post-war examples of design and development collaboration are the Sohio–Badger acrylonitrile process and the ICI–Kellogg ammonia process. It is likely that the large oil and chemical companies will remain the major source of new processes, because of the scale of their R & D, and their experience in plant operations. Advances in science might conceivably reverse this trend to some extent. Pilot-plant work in the development of new processes is still often necessary because scientific knowledge of the probable behaviour of liquids and gases is still not sufficiently precise to be able to predict with complete assurance the effects of scaling up. In so far as the growing precision of scientific knowledge and the use of computers for design calculations make it possible to eliminate the pilot plant stage in process development, this could redress the balance in favour of smaller firms and R & D groups. However, the complexity of the design problems would still militate in favour of large R & D laboratories. The relevance of these R & D scale problems to size of firm is discussed more fully in chapter 6.

Whilst corporate R & D has come to dominate the major revolutionary leaps in technology, it is important not to overlook the steady growth in productivity brought about by relatively minor technical improvements to existing processes. Enos's account of technical innovation in the oil industry confirms the results of Hollander's (1965) very detailed study of technical change in Du Pont rayon plants on this point. Hollander shows that the greater part of productivity increase was attributable to 'minor' technical advances introduced largely as a result of the activities of the Engineering Department and Technical Assistance Groups. The process of technical change in industry thus takes two main forms: radical innovations in products and processes which have increasingly originated in professional R & D laboratories in universities, industry and government; secondly, incremental improvement of products and processes associated with increasing scale of investment and learning from experience of production and use. However, it would be dangerous to conclude that scale economies have reached their limits in these or other industries. In his interesting discussion of this problem, Gold (1979) points to the fact that the Japanese doubled the size of blast furnaces beyond the established 'optimum' level.

Nuclear energy

The brief account of process innovation given here has stressed the shift from the inventor–entrepreneur of the nineteenth century towards large-scale corporate R & D. This differs sharply from the interpretation given by Jewkes and his colleagues in their classic study of *The Sources of Invention* (Jewkes, Sawers and Stillerman, 1958). They minimize the differences between the nineteenth and twentieth centuries and generally belittle the contribution of corporate professional R & D. They argue that most important twentieth-century inventions were the result of the work of individual inventors as in the nineteenth century, either free-lancing or working in universities. Part of this difference in emphasis is due to the fact that Jewkes was concerned primarily with *invention*, whereas this account is concerned primarily with *innovation*. Jewkes and his colleagues concede that the costs of *development* are often so high that large-scale corporate R & D may be necessary to bring an invention to the point of commercial application. Of the forty major twentieth-century inventions which they attribute to individual inventors (compared with twenty-four attributed to corporate R & D) at least half, according to their own account, owed their successful commercial introduction to the development work and innovative efforts of large firms.

Thus Jewkes emphasizes most strongly the contribution of *Houdry* as an individual inventor, whereas this account has emphasized his collaboration with the oil companies in the later stages of his work, and the even heavier costs of the fluid catalytic-cracking process, developed almost entirely through corporate R & D. However, it is important to note that Enos supports Jewkes' view of the origin of *inventions* in the oil refining industry.

It can reasonably be maintained that, from the standpoint of economics, it is *innovation* that is of central interest, rather than invention. This is not to deny the importance of invention, nor the vital contribution of creative individuals to both invention and innovation. There is no inconsistency between Jewkes' emphasis on the importance of university research and invention and the interpretation which is given here. Nor is it denied that the lone wolf and the inventor-entrepreneur still play an important part. (This is discussed at greater length in chapters 6 and 7.) But here too, even on Jewkes' own account of major inventions, there has been a shift over time since the early part of the twentieth century towards a larger contribution from inventors associated with corporate R & D. Moreover, he conceded that this contribution has been particularly important in the chemical industry as compared, say, with mechanical engineering. If, as is argued here, the chemical 'pattern' is gradually becoming more typical of industry as a whole, then the differences between nineteenth- and twentieth-century invention and innovation cannot be so lightly dismissed.

Not surprisingly, in view of the difference in approach, Jewkes and his colleagues dismiss nuclear power somewhat contemptuously and do not even count it in their list of important twentieth-century inventions (although they include rockets, cinerama, the self-winding wrist watch, and the zip fastener). The omission is deliberate and is explained on the grounds that the subject is too large and that it is still surrounded by too much secrecy. They also justifiably pour much cold water on the over-optimistic estimates of the economic advantages of nuclear power.

It is perfectly true that some of the early hopes associated with nuclear power have not been fulfilled and in particular that there has been a major problem of cost escalation and disappointingly low load factors with most types of reactor, except the Canadian CANDU (Surrey and Thomas, 1980). There have also been other legitimate public anxieties concerned not so much with the past safety record (which has been extremely good) as with the possibility that the rapid world-wide diffusion of civil nuclear power may increase the dangers of nuclear weapons proliferation and the ultimate possibility of very serious accidents arising from human errors. These fears and anxieties cannot simply be dismissed out of hand and, as Cottrell (1981) has pointed out, the various Atomic Energy Authorities have often been their own worst enemies through their own insensitivity, secrecy and illegitimate over-simplification.

Nevertheless, it can hardly be denied that nuclear power does represent an extremely important twentieth-century innovation and that it has at least to some degree relieved the constraint on energy supplies arising from the limits on fossil fuels and the delays in developing other safer and hopefully cheaper energy technologies and energy conservation measures. It is also of interest here because it represents in many ways an extreme manifestation of some of the trends which have been discussed in relation to chemical process plant and oil refineries. The development of several types of nuclear reactor is the *culmination* of the tendencies towards large-scale professionalized R & D characteristic of the twentieth century. It may be that the diseconomies of scale which have already been discussed and the related international political and environmental problems will bring the further development of nuclear power to a halt, as has already occurred in Austria (Lönnroth and Walker, 1979). But it seems more likely that it will continue to play an important role in some countries until it is superseded as a result of further technical innovation in various alternative energy systems. These might include fusion power, which at present seems still some way off, but would represent a further giant step down the road of large-scale, government-financed R & D and massive centralized investment. Present assessments seem to indicate that it would be a much safer technology than the contemporary generation of nuclear reactors or the fast-breeder reactors under development.

It would be a reasonable objection to the very high priority accorded to nuclear power in public allocations for R & D that it pre-empted almost all the available resources for new energy technologies. The strength of the political lobbies associated with the centralized powerful atomic energy authorities was an important factor in this outcome and is further discussed in chapter 9.

Public involvement in the civil development of nuclear power arose from the government's role in weapons development and Jewkes may be right in believing that military security slowed down the early civil work on reactors for power systems, increased its cost and inhibited the participation of some groups in industry and universities who might have been able to make a fruitful and critical contribution. He may also be right in believing that some public nuclear R & D programmes were unnecessarily lavish because of the lack of normal commercial constraints. Duncan Burn (1967) has provided strong supporting evidence.

However, it is difficult to see how reactors could have been developed except at very high cost and without a considerable degree of government participation in financing the R & D. In fact, nuclear reactors have not been developed anywhere in the world without such public sector involvement. The development costs of the principal reactor systems as far as they were known up to the end of 1970 are shown in Table 2.6. From this it is apparent that reactor development has been much more expensive than the most complex chemical processes. The new fast breeder reactors under development in the 1970s and 1980s are proving even more dauntingly expensive, so that the £124 million 'projected additional' expenditure of 1971 has now been dwarfed. A very disquieting feature of the R & D on fast-breeder reactors has been the large-scale systematic and persistent under-estimation of real development costs which has characterized all the main projects (Keck, 1977, 1980 and 1982).

The need for such massive public investment arose because nuclear engineering processes carry to an extreme degree all the tendencies which have been discussed in relation to chemical and oil refinery processes. The very heavy costs and long gestation period arose from the extraordinary complexity of the design problems, involving new materials, instruments, components and equipment of all kinds to satisfy the exacting requirements and safety standards of the new technology.

Table 2.6 Development costs of various types of reactor as estimated by the Atomic Energy Authority, 1971 (£ million)

Reactor type	Actual	Projected additional
Gas-cooled, Magnox mark 1	20	
Gas-cooled, AGR mark 2	114	
High-temperature reactor, mark 3	25	32
Steam-generating heavy water reactor	78	16
Fast-breeder reactor	205	124
Total	442	172

Source: UKAEA (1971).

At every stage intimate collaboration was necessary between nuclear engineers and scientific research teams investigating fundamental problems, so that large R & D groups were essential.

The whole of the Manhattan Project on nuclear weapons and the subsequent projects on civil applications of nuclear energy were based on nuclear physics. They were totally science-dependent and could in no way have emerged from existing practice or incremental improvements in conventional methods of generating electricity or making explosives. Whilst this background research is itself extremely expensive, it is dwarfed by the huge costs of prototype and pilot plant work for new types of reactor. Experience has shown that only full-scale plant operation has been able to resolve some of the complex technical problems and one of the errors has been to move too quickly from small-scale experimental reactors to batch orders for several reactors of an unproven design (Rush, Mackerron and Surrey, 1977).

3 Synthetic Materials

Plastics, although almost entirely a twentieth-century industry, are already one of the world's main groups of industrial materials. World plastics consumption by weight already exceeds that of non-ferrous metals and in volume terms it is far greater. Synthetic rubber overtook natural rubber consumption in the 1960s, and man-made fibres already accounted for nearly half of total fibre consumption by 1980 (Table 3.1).

Table 3.1 World production of various materials

Material	Volume of world production[a] (million metric tons)					
	1913	1938	1950	1960	1970	1980[b]
plastics	0.04	0.3	1.5	5.7	27.0	40.0
aluminium[e]	0.7	0.5	1.3	3.6	8.1	11.2
zinc[e]	0.8[d]	1.4	1.8	2.4	4.0	4.8
copper[e]	1.0	1.8	2.3	3.7	6.1	..
steel	53[e]	88	153	241	448	480
synthetic rubber	−	0.01	0.5	1.9	4.5	7.7
natural rubber	0.12	0.92	1.9	2.0	2.9	3.7
synthetic fibres[f]	−	−	0.12	0.65	4.5	8.4
cotton, lint	..	5.2	6.0[g]	7.1	7.7	9.1
wool, raw	..	1.6	1.7	2.1	2.2	2.2

− Nil.
.. Not available.
[a] excluding USSR, China, Eastern Europe.
[b] provisional estimates for 1980.
[c] primary refined production.
[d] zinc 1909.
[e] steel 1910.
[f] pure synthetics, i.e. excluding rayon.
[g] cotton 1951.
Source: UN Yearbook of Statistics, New York; Saechtling (1961); *UN Monthly Bulletin of Statistics*; author's estimates.

Their growth rate has been extremely high, largely because they have outstanding technical and cost advantages in a wide range of applications, and because of actual and anticipated shortages of naturally occurring materials. From the 1940s to the 1970s, competitive substitution for older materials played a big part in the very high rates of growth of production and consumption; inevitably, as they came to account for a high proportion of the combined total consumption, their rate of growth slowed down in the 1970s and began to asymptote towards the slower rate of the natural materials. Plastics are light, easy to fabricate and install, frequently have good electrical insulation and excellent resistance to corrosion and pests. With synthetic rubbers they can increasingly be tailor-made or blended to meet the requirements of any particular application, but they have the disadvantage of the rather limited temperature range within which most of them can be used. Some of the newer plastic materials can be used at very high temperatures and are extremely strong, but are still relatively expensive. Synthetic fibres also have properties of strength, durability and resistance to pests, which in many cases surpass or supplement those of natural fibres, greatly enlarging the range of possibilities for the textile industry. Paradoxically their very virtues lead to problems of waste disposal and pollution.

As they are usually defined, synthetic materials differ from similar older man-made materials, such as glass and ceramics, in their organic origin. They are composed of giant molecules of organic substances based on long chains of carbon atoms. For casein and cellulosics, such as rayon, these chains or polymers are of natural origin, but the vast majority of the newer materials are synthesized from simple chemical units or monomers, such as ethylene (polyethylene) or styrene (polystyrene). These polymers can be made to flow, on the application of adequate heat and pressure, to any desired shape, which is maintained when the heat or pressure is withdrawn. Thus the same basic material, such as nylon, or polypropylene, may be used as a fibre, or as a sheet, or film, or moulded to form a component or product of a specific shape. The fundamental chemical knowledge required to manufacture and blend the true synthetics is much greater than for the cellulosics, which were the nineteenth-century plastics innovations. Advances in plastics technology have in turn strongly influenced metallurgy, so that the new 'materials science' has developed embracing both.

The major synthetic materials are manufactured by the chemical industry, using raw materials such as petroleum products, natural gas and coal, or intermediates derived from these materials such as ethylene, propylene or acetylene. The basic chemical producers may deliver the materials in the form of solid or liquid resins, moulding or extrusion compounds or emulsions to the fabricating industry, just as the metal producers deliver metal to the engineering industry. Alternatively, the chemical producers may turn out film, sheet, fibre, rods, tubes and other mouldings and extrusions, or fabricate more complex products themselves. The extent of vertical integration varies but because of their know-how and the economics of integrated production, chemical firms have become increasingly involved in the textile industry, and to a lesser extent in the manufacture of building and packaging materials and engineering components. Another important factor which compelled the chemical firms to enter these industries was the need to develop new markets for their materials in areas which were resistant to technical change or unwilling to experiment with new materials. Although their advantages are now widely recognized, in the early days of synthetic materials they were treated with almost universal scepticism. Frequently it was the shortage and high price of natural materials or the prospect of war which led to experiments with new materials and ultimately to innovations.

The early synthetic materials and the inventor–entrepreneurs

Of the sixty or so major synthetic materials (plastics, rubbers and fibres), all except about a dozen were innovated by large chemical firms (see Freeman, Young and Fuller, 1963; Hufbauer, 1966). These exceptions were mainly the earlier innovations which were usually brought to the point of commercial application by inventor–entrepreneurs, such as Baekeland (bakelite), Chardonnet (rayon), Parkes and Hyatt (celluloid). The professionalization of R & D and the role of the industrial R & D laboratory is clearly apparent as we move into the twentieth century. None of the synthetic materials was invented, developed or innovated by the suppliers or fabricators of the natural materials and metals for which they were largely substituted. But after the new materials had been introduced a few of these firms took an interest in them, made improvements and developed new applications. In the case of fibres and rubbers, some of the principal users of natural materials, such as tyre-makers, played a similar role.

Perhaps the man who had the strongest claim to be the original inventor of the first true synthetic was the British chemist Alexander Parkes, who called his material 'Parkesine'. Patented in 1865, it was made from cellulose nitrate and oils and was the forerunner of *celluloid*. But the Parkesine Company which Parkes established to exploit his invention went bankrupt. The combs and other products which he sold were defective, because he had failed to solve the problem of suitable plasticizers and solvents satisfactorily. Although he was not formally trained as a chemist, Hyatt successfully used camphor as a plasticizer so that his company in the United States established celluloid as the first commercially viable plastic material. This illustrates at the very outset the international character of the inventive process in this industry and its risks.[1]

In a similar way in the rayon industry, a series of British inventions solved some of the critical problems of inflammability which caused the French inventor-entrepreneur, Count Chardonnet, to suspend production of his new *nitrocellulose rayon* in Besançon. Like Parkes, Chardonnet could claim to be a pioneer, but he too launched production (in 1884) before his product was in a satisfactory state. The fibre was weak and brittle and the available textile machinery was not adapted to it. However, Chardonnet had the resources to persist and was able to resume production in France and Germany in the 1890s. But *viscose rayon* and *cellulose acetate* later proved far more successful than the original nitrocellulose rayon. The viscose process was the invention of a British consulting chemist, C. G. Cross, in 1892, but it was some time before a spinning technique was developed.

Another early plastic material to be pioneered by a new firm was '*Galalith*' made from casein and formaldehyde and widely used for buttons. The German chemist, Spitteler, collaborated with Krishe, a businessman, in setting up the successful International Galalith Company (Vereinigte Gummiwaren) in 1899.

Although the first innovations in the rayon industry were made by inventor-entrepreneurs, it was not long before much larger firms with greater financial resources became involved. The Viscose Spinning Syndicate which had been formed in 1900 to exploit Cross and Bevan's viscose rayon process, sold out to Courtauld's, the dominant firm in the British silk industry, in 1904. From then onwards Courtauld's held a monopolistic position in the British rayon industry and later a very powerful position in the world industry.

Similarly, Hyatt's dental-plate company in Albany, which first made a commercial success of celluloid, was soon taken over by a much stronger financial grouping. Brandenberger, the Swiss-born French chemist who invented cellophane film and took out world patents in 1912, came to an arrangement with the French rayon cartel to form a new subsidiary, known as 'La Cellophane', to manufacture and sell his product. It came on the market in 1917. Thus a new pattern began to emerge in the twentieth-century industry, in which the role of the inventor-entrepreneurs became less significant, and the large rayon firms dominated product and process improvement.

[1] On the early history of the plastics industry see Kaufman (1963); Hufbauer (1966); Yarsley and Couzens (1956); see also de Bell (1946).

However, the most successful of the early inventor–entrepreneurs was undoubtedly Leo Baekeland, the Belgian chemistry professor, who set up his own firm in the United States in 1910 to manufacture the condensation plastic which bore his name. He had previously worked for several years as a private inventor and in American industrial research. Not only did he take out the original patents for 'Bakelite', a phenol-formaldehyde resin, he also personally pioneered most of the early applications. He built up a very successful commercial enterprise with subsidiaries and licensees in many countries. Although it accounts today for less than 5 per cent of total plastics consumption, Bakelite was the most important synthetic resin in the inter-war period.

The main synthetic materials

The synthetic rubbers and almost all of the other major synthetic materials and fibres introduced after the First World War were innovated by established large chemical firms, with extensive research and experimental development facilities. Very often, of course, fundamental chemical discoveries and inventions were made in university laboratories, and in particular Staudinger's work at Freiburg on long chain molecules provided the theoretical basis for many of the industrial advances of the 1930s. But years of intensive applied research and pilot plant work have usually been necessary to take a material from the stage of a laboratory curiosity to that of a commercially viable process for a reasonably homogeneous and stable product. In addition, a great deal of applied research and experimental development work has been necessary to explore the vast number of potential applications, and to modify the materials to create the variety of grades or blends to suit each particular end-use.

Noreen Cooray (1980), in her study of the substitution of synthetic rubbers for natural rubber, found that one of the main comparative advantages of the synthetics was the breadth and depth of the applications research, which meant that there were 'tailor-made' modifications, blends and specialities for a great variety of specific requirements. She came to the conclusion that natural rubber could only compete effectively if the producers of natural rubber organized a comparable applications research programme and did not confine themselves simply to improving the techniques of cultivation, even though that type of research had been very successful in producing high-yielding varieties of tree and reducing costs and prices of natural rubber.

To some extent the advantages of polymer R & D and applications research were cumulative, reinforcing the technical leadership of the strongest industrial laboratories. The early inventor–entrepreneurs usually financed their own experiments and took out their own patents before establishing an innovating firm. The pattern changed completely with the newer synthetics. An example which indicates this change very clearly is that of 'Terylene', the ICI polyester fibre (known under the trade name of 'Dacron' in the US and 'Trevira' in Germany). This fibre was invented in 1940 in the small R & D laboratory of a moderate-sized textile firm, the Calico Printers' Association. But Calico Printers were unable to develop the process and innovate the fibre. It was licensed to Du Pont in the United States and to ICI in Britain, who were able to bear the extremely heavy expense of pilot plant work, applications research, trial production and trial marketing of the new fibre. It was estimated that the total costs were of the order of £10 million for ICI and a similar sum for Du Pont, before they were able to market the product in 1950.

In many other cases both the inventions and the development took place in the R & D laboratories of large chemical firms, as with nylon, polyethylene, PVC, and Corfam, all of which are briefly described at the end of this chapter. But even where the original discovery or invention was made in a university (as with neoprene and methyl methacrylate) or by a smaller firm (as with Terylene), the costs of development work, the problems of marketing and the scale of investment in new plant led to the actual innovation being launched by the larger firms.

Among the firms which have an outstanding record in technical innovation in synthetic materials are IG Farben, Du Pont and ICI, the largest chemical firms in Germany, US and Britain respectively. Not only did these firms account for a number of the most important synthetic innovations themselves (PVC, nylon, polyethylene, polyesters, acrylics, polystyrene, buna, neoprene, etc.); they were usually among the first 'imitators' or 'adopters' of innovations made by others and played a considerable part in developing new machinery and a wide variety of new applications. One of the major innovative achievements of each of them is discussed, but first the overall contribution of the German chemical firms is reviewed.

Research and experimental development at IG Farben

As we have seen in chapter 2, German chemical firms had already established a strong tradition of generating their own new products and processes during the nineteenth century, building up the synthetic dyestuffs industry to become world leaders on this basis. Thus they were already accustomed to heavy long-term investment in R & D programmes long before they formed the IG Farben trust in 1925. Regular statistics of expenditures on R & D were not kept in any country until the 1950s, but some firms who pioneered professional R & D kept records for earlier periods, and fortunately we have such figures for IG Farben (Ter Meer, 1953). It is true that these estimates are not completely consistent with modern definitions[2] but discussion with German chemists suggests that these differences are small.

From 1925 to 1939 it is fairly clear that, in absolute terms, the R & D activity in IG Farben in synthetics was far greater than in any other firm, and indeed its total R & D programme was the biggest in the world. IG Farben's total R & D expenditure averaged just over 7 per cent of turnover from 1925 to 1939. This is a higher ratio than in most large chemical firms since the war, which are mainly in the range of 3 to 5 per cent. From 1925 to 1931, IG's expenditure was between 7 and 10 per cent, but in the world recession it was cut back fairly drastically to 4.9 per cent in 1933. It is notable that the main economies were in development expenditure, and that the research staff were maintained throughout at more than 1,000 qualified scientists and engineers. From 1934 to 1939 research and experimental development expenditure rose again to between 5 and 6 per cent of turnover. These figures exclude technical services and extra-mural grants and

---

[2] For definition of R & D see the Appendix. 'Research-intensity' of a firm or an industry may best be measured by relating R & D expenditures to net output. This enables satisfactory inter-firm and inter-industry comparison because it adjusts for the differences in value of bought-in materials and components. However, since figures for net output are seldom available, frequently the less satisfactory measure of R & D expenditure as a ratio of turnover has to be used, or of R & D manpower as a ratio of total employment. In the case of IG Farben the figures available are for R & D as a ratio of turnover. The net output ratio would probably be nearly twice as high, but cannot be exactly estimated.

donations to universities (which are sometimes included with R & D to give a misleadingly inflated figure) and exclude capital expenditure on new laboratories and instruments etc. This would normally add about 1 percentage point to the figures, making them even more impressive. Throughout this period the firm spent more on research than it distributed in dividends. According to one account, it was the need to concentrate research resources and make large investments in high polymer chemistry that finally persuaded the constituent firms to come together in 1925 to form the trust. It is an indication of the high priority given to research that there was no pruning before the world depression, but a substantial increase from an already high level of expenditure.

Another notable feature of IG's pre-war R & D was the importance attached to close contact and cooperation with fundamental researchers in the universities and other academic institutes. This continued the approach of the constituent firms, Bayer, BASF and Hoechst, all of whom had a tradition of employing outstanding academic consultants, including a number of Nobel prize winners. They funded a great deal of research in German universities and tried to create conditions in their own laboratories which would attract the best chemists. Both the management and the R & D departments were dominated by graduate chemists. In the case of synthetic materials, the outstanding world authority on macromolecular chemistry, Professor Staudinger of Freiburg, was an active consultant with IG throughout the inter-war period, and it is notable that Ziegler, who made the major theoretical contribution to the post-war plastics industry, was already a consultant to Hoechst when he made the critical discoveries at Mülheim which led to the innovation of low-pressure polyethylene in the 1950s. A new consultancy agreement was then signed which covered licensing arrangements for his process.

Finally, we may note that IG Farben sometimes followed a strategy of 'parallel teams' in experimental development work. No doubt this was partly due to the local pressures of the constituent firms, but it was also a deliberate strategy of the coordinating research apparatus, which adopted a philosophy of experimenting with several alternative routes in developing processes for the new synthetic materials.

Patents as a measure of inventive output

We may now turn to the 'output' of IG Farben's research and experimental development. The scale of their R & D was undoubtedly much larger than that of any other privately-owned firm in the world in the inter-war period. But how effective was it? In order to answer this question we must first spend some time discussing patents and other output measures.

The measurement of efficiency in R & D is one of the most complex problems in management economics, and there is no simple answer to this question. 'Inputs' into R & D can be measured and reduced to a common financial denominator, but even here there are serious complications, such as the attribution of information inputs from outside the formal R & D structure. But when it comes to the measurement of 'output' the difficulties are overwhelming. They are discussed in some detail in UNESCO (1970). Here it is only possible to indicate two complementary ways of approaching the measurement of 'effectiveness' of IG Farben's R & D—in terms of numbers of patents and numbers of innovations. It is not suggested that either of these is satisfactory or would be employed if better information were available. But taken together they do enable us to give some kind of answer on the basis of 'cost-effectiveness', although not on the basis of 'profitability' of innovations. To assess the 'profitability' of IG Farben's R & D we would need far more detailed information about the costs and profit margins on their new products than is ever likely to be published. There is also the enormous complicating factor of the German war economy.

Patents are a measure of *inventive* output rather than innovative success, and therefore should be used together with some measure of innovation. But provided their limitations are kept in mind they are probably much more useful than is commonly believed in industry. Schmookler did more than any other economist to demonstrate their value in economic history and he concluded from his extraordinarily thorough studies of inventions in the railroad industry, petroleum refining, paper-making and construction that patent statistics provided a more satisfactory indicator of 'inventive output' in the United States from 1850 to 1950 than lists of 'important inventions'. In his view, they reflected all the minor and improvement inventions and avoided the bias inherent in any subjective assessment of 'importance' (Schmookler, 1966). While accepting Schmookler's point about the value of aggregate patent statistics as a measure of incremental invention, we cannot follow him on the question of key patents as a measure of radical inventions.

One of the limitations of patent statistics is the variation between industries and firms in 'propensity' to patent. For example, in defence-oriented industries there is usually a lower propensity to patent, and this might account, to some extent, for the much lower ratio of patents per unit of R & D expenditure in aircraft and electronics (Table 3.2a and Table 3.2b).[3] However, in the chemical industry there is a high propensity to patent; it is quite exceptional to find a chemical firm which does not attach importance to securing patent protection for its inventions, and it is difficult to identify any major technical advances in the plastics industry which were not the subject of patenting activity.

It is generally agreed by those who have attempted to use these statistics that annual fluctuations and variations in quality are such that it makes more sense to analyse the figures for groups of five or ten years. Another complicating factor arises from the variations in national patent legislation.[4] To overcome these difficulties, Pavitt and Soete (1980) have pioneered the use of statistics of foreign patents taken out in the United States, which puts all countries (except the US) on a similar basis. As the US is the most important single market, it is reasonable to assume that firms will register their important patents in that country and they have demonstrated extremely interesting results on the relationship between foreign trade performance and patents taken out in the United States (Soete, 1981 and Pavitt, 1982).

[3] There are, of course, other reasons for these differences. The large number of patents in instruments and mechanical engineering relative to R & D is partly due to the fact that invention in these industries is still conducted to a significant extent outside the formal industrial R & D system. There are also classification problems for the two series of statistics.

[4] In some countries patents are granted without examination for originality, whilst in others there is an examination procedure. This means that in countries such as France, Belgium and Italy about 90 per cent of patent applications lead to the grant of a patent, but the proportion falls to about 60 per cent in the US and UK, and still lower in Germany, Netherlands and Scandinavia. An international comparison would thus tend to be biased against German firms a comparison with British and American firms, and still more with French. The most important patents will be taken out in all the major manufacturing countries, but the statistics for any one country will normally be biased towards the firms domiciled in that country (including foreign-owned subsidiaries operating there).

Table 3.2a Patents delivered in various branches of British and French industry compared with research expenditure, 1961

Industry	Percentage of total number of patents delivered		Percentage of total research expenditure, manufacturing industry			
	United Kingdom	France	United Kingdom		France[a]	
			Excluding aircraft	Including aircraft	Excluding aircraft	Including aircraft
Aircraft	1.7	1.8	–	38.4	–	32.7
Electrical engineering and electronics	22.2	17.3	38.3	23.5	35.0	23.6
Instruments	6.3	10.6	4.1	2.5	0.7	0.5
Chemicals and oil products	24.0	20.6	20.3	12.5	27.5	18.4
Vehicles	5.0	6.6	4.3	2.7	11.1	7.5
Engineering	18.0	16.3	13.1	8.0	9.8	6.6
Metals and metal products	9.5	7.5	6.1	3.7	5.1	3.4
Building materials, wood and furniture, building	6.4	11.0	6.0	3.7	5.4	3.6
Textiles and clothing	5.4	6.3	3.8	2.4	3.6	2.4
Food, drink and tobacco	1.5	1.9	3.2	2.0	1.8	1.2
Total	100	100	100	100	100	100

[a] France, 1966.

Source: Fabian 1963; *Report of the Comptroller General of Patents,* 1961; *Bulletin de la Propriété Industrielle-Statistiques,* 1961; *Report of Advisory Council on Scientific Policy,* Cmnd. 1920, 1963; 'Les moyens consacrés à la recherche et au développement dans l'industrie fançaise en 1966' (1968), *Le Progrès Scientifique,* Numéro Spécial.

Table 3.2b Sectoral shares of US industrial patenting compared to other activities

Sector	Patents granted (1973) (%)	Total R & D expenditures (1973) (%)	Industry-financed R & D (1974) (%)	Qualified scientists and engineers (1975) (%)	Manufacturing employment (1974) (%)	Manufacturing sales (1974) (%)
Food and kindred products	1.0	1.2	2.1	3.5	6.9	10.2
Textiles	0.9	0.3	n.a.	1.2	3.3	1.8
Chemicals (except drugs)	11.3	6.7	15.6 }	13.1	7.9	9.6
Drugs	1.3	2.9		3.1	3.4	(14.9)
Petroleum and related products	1.3	2.3	4.4	3.0	2.7	2.3
Rubber	4.1	1.3	n.a.	2.3	2.5	1.6
Stone, glass, clay, concrete	1.9	0.8	1.3 }	4.9 }	7.8 }	7.8 }
Ferrous metals	0.7	0.7				
Non-ferrous metals	0.6	0.6	2.2	7.3	5.1	3.4
Fabricated metals products	11.2	1.3	2.0	12.5	11.2	8.5
Non-electrical machinery	25.9	10.2	15.3	21.5 }	16.6 }	10.6 }
Electrical and electronics	20.0	30.6	21.2			
Motor vehicles	2.8	11.3	23.8 }	18.5 }	16.5 }	14.7 }
Aerospace	1.8	23.6				
Scientific instruments	9.8	4.2	5.8	4.9	3.6	2.5
Other manufacturing	5.5	2.0	6.4	4.2	12.5	14.1
Total	100.0	100.0	100.0	100.0	100.0	100.0

Source: K. Pavitt, 'R & D, Patenting and Innovative Activities: A Statistical Exploration', Research Policy, Vol. 11, No. 1, February 1982, pp. 33–51.

After this brief discussion of some of the problems of patent statistics, we may now return to consideration of the role of IG Farben in the development of plastics. Fortunately, in the case of synthetic materials a three-volume classified anthology of all international patents has been painstakingly assembled by Delorme (1962) for the entire period, 1791–1955. This avoids double counting and any bias which exists in the statistics derived from this source is probably towards French and American rather than German firms. The outstanding feature of an analysis of these figures is the dominant position of IG Farben in the period from 1931 to 1945, and of its predeccessors before 1930. The firm accounted for over a third of all patents taken out by the thirty largest firms (Table 3.3), and since large firms accounted for a high proportion of all patents this was equivalent to 17 per cent of the world total from all sources. No other firm except Du Pont has as many as one sixth of the number of patents taken out by IG in this period. In the field of vinyl patents it accounted for over a quarter of all world patents. Between 1925 and 1930 IG registered twice as many plastics patents as any other firm in the world for the whole period from 1791 to 1930.

Altogether more patents were taken out in plastics during the fourteen years 1931–45 than in the previous 140 years, and a notable feature of the long-term trend was the decline in patents taken out by individuals by comparison with corporate patents (Table 3.4). In the post-war period patents awarded to individuals had declined to less than 10 per cent of the total. This reflects the increased contribution of professional R & D organizations within firms by comparison with the private inventor, and the increasingly science-based character of the inventive process in this industry.

In the period up to the Second World War, German and American firms were responsible for over 80 per cent of all patents taken out by firms but, with the exception of Du Pont, the leading US chemical firms came relatively late into this field. The early plastics manufacturers, Bakelite and Celluloid, were patenting on a significant scale, as were Eastman Kodak from the film side, and General Electric, one of the most research-intensive American firms with very wide interests in new materials. Both played an important part in the earlier period and are still among the leaders, for example in polyacetals and polycarbonates. Röhm and Haas was at this time based mainly in Germany and was principally concerned with the development of acrylic materials (methyl methacrylate). By the 1930s, ICI was among the leaders although still behind IG Farben in the range of its plastics research and production. Over a long period, the highly research-intensive Swiss chemical firm, CIBA, has been consistently among the leading firms in numbers of patents taken out and has several major innovations to its credit, notably in epoxy resins. But with the exception of ICA and CIBA, the remainder of the European chemical industry was far behind the German and American firms.

From 1946 to 1952, IG Farben was being reorganized by the Allied Military Governments and was not in a position to take out any patents. Moreover, many of its secrets were compulsorily made available to British, French and American firms in 1945–6 by Allied investigation teams. It was not until 1952 that the successor firms to the dissolved combine were able to resume normal production and research activity. Consequently, the patent statistics for 1946–55 show American firms in a dominant position with eight of the ten leading firms, and Du Pont as the established world leader. However, the combined total of the successor firms to IG Farben, even in this period, was greater than that of any other firm except Du Pont.

Table 3.3   Patents for plastic materials taken out by leading firms

30 leading firms' patents taken out in UK, US, France, Germany

1791–1930	No.	1931–45	No.	1946–55	No.
1 IG Farben	346	1 IG Farben	889	1 Du Pont	637
2 Eastman Kodak	169	2 Du Pont	321	2 Monsanto	283
3 Du Pont	78	3 Röhm and Haas[c]	145	3 American Cyanamid	266
4 Celluloid	66	4 Hercules Powder	132		
5 Bakelite Corp	59	5 GE	120	4 Shell/N.V. de Bataaf	263
6 Bayer[a]	55	Eastman Kodak	120		
Meister, Lucius and Brüning[a]	55	7 Dow	115	5 ICI	253
		Kodak-Pathe and Kodak	115	6 Röhm and Haas[c]	210
8 CIBA	42			7 Dow	187
9 Bakelite GmbH	40	9 ICI	90	8 B. F. Goodrich	160
10 BASF[a]	38	10 Carbide and Carbon	88	9 US Rubber	156
GE	38	11 Phrix Arbeitsge-meinschaft	73	10 Eastman Kodak	140
12 British Thomson-Houston	35	12 Celanese	67	11 Standard Oil/Esso	131
13 Consortium für Elek[b]	33	13 A. Wacker Ges.	65	12 BASF[a]	115
14 American Cyanamid	28	14 American Cyanamid	60	13 F. Bayer[a]	111
15 British Celanese	26	15 CIBA	56	14 CIBA	101
		16 Ellis-Foster	51	15 St. Gobain	77
				16 Distillers	74

30 leading firms' patents taken out in UK only

1954–8	No.	1959–62	No.
1 ICI	299	1 ICI	485
2 Du Pont	288	2 Du Pont	428
3 Standard Oil/Esso	243	3 F. Bayer[a]	346
4 F. Bayer	199	4 Union Carbide	327
5 US Rubber	170	5 Standard Oil/Esso	246
6 Midland Silicones	168	6 Hoechst[a]	207
7 Monsanto	143	7 Montecatini	205
8 GE	128	8 CIBA	168
9 Celanese	113	9 Dow Chemical	162
10 Courtaulds	109	10 Phillips Petroleum	153
11 Shell/NV de Bataaf.	108	11 BASF[a]	145
12 Union Carbide	107	12 Röhm and Haas	132
13 BASF[a]	106	13 Shell/NV de Bataaf.	118
14 Dow Chemical	100	14 American Cyanamid	112
15 Röhm and Haas	87	15 US Rubber	110
16 American Cyanamid	76	16 Midland Silicones	109

Company	n
16 Barrett	25
Ellis–Foster	25
ICI	25
19 Cie. Fr. Thomson–Houston	24
Naugatuck	24
21 Kroll	23
Canadian Electric Pathe	23
24 AG für AF	21
Kunstharz Pollak	21
26 Chem. Fab. Griesheim	16
27 Carbide and Carbon	15
PF Instruments	15
29 E. Schering	14
30 Hercules Powder	13
Soc. Chem. des Usines du Rhône	13

Company	n
17 Bakelite	48
18 Deutsche Hydrierwerke	46
19 Pittsburg Plate Gl.	44
20 British Celanese	42
21 Standard Oil	41
22 Bakelite Ges.	40
Cie. Fr. Thomson–Houston	40
24 Monsanto	37
25 Deutsche Kelluloid	35
26 B. F. Goodrich	32
27 Thuringische Zellwolle	30
28 Rhône-Poulenc	26
29 Harvel Research	24
30 Chem. des Elek[b]	24

Company	n
17 Gen. Aniline and Film	73
Celanese	73
19 Wingfoot	72
20 Cie. Fr. Thomson–Houston	69
Carbide and Carbon	69
22 Hercules Powder	65
23 Hoechst[a]	57
24 Phillips Petroleum	55
25 Kodak–Pathe	47
26 American Viscose	45
27 Chemstrand	44
28 Rhône-Poulenc	43
GE	43
30 Chem. Werke Hüls[a]	42
Koppers	42

Company	n
Dow Corning	76
18 B. F. Goodrich	75
19 Hoechst[a]	68
20 Hercules Powder	64
Distillers	64
22 Dunlop Rubber	60
23 Phillips Petroleum	57
24 CIBA	56
25 Wingfoot	38
26 Minnesota Mining	28
27 Chemstrand	23
Rhône-Poulenc	23
Wacker–Chemie	23
30 Gen. Aniline and Film	20

Company	n
17 Monsanto	100
18 GE	92
19 B. F. Goodrich	83
20 Dunlop Rubber	79
21 Minnesota Mining	77
22 Courtaulds	73
23 Celanese	60
24 Hercules Powder	63
25 Distillers	53
26 Chemstrand	51
27 Rhône-Poulenc	47
28 Chem. Werke Hüls[a]	42
29 Wacker–Chemie	41
30 Dow Corning	34

a Part of IG Farben.
b Undertaking research in association with A. Wacker.
c The German and American parts of Röhm and Haas are listed together here.
Source: Delorme (1962) and Patent Office, London.

Table 3.4   Patents issued for the principal groups of plastics, 1791–1955

Patents taken out by	1791–1930		1931–45		1946–55	
	Number	Per cent of total	Number	Per cent of total	Number	Per cent of total
Individuals	1803	43	791	15	489	8
Firms	2436	57	4341	85	5749	92
Total	4239	100	5132	100	6238	100

Source: Delorme (1962).

Unfortunately, Delorme's anthology does not extend beyond 1955, but using British Patent Office statistics it is clearly apparent that, in the 1950s and 1960s, the leading German chemical firms continued their recovery (Table 3.3) although not achieving the pre-war dominance of IG Farben. British national patent statistics are biased in favour of British domiciled firms but probably not biased in favour of German as against American firms or vice versa, except possibly in the case of Monsanto and Standard Oil, which had major subsidiaries operating in the UK.

Because of the importance of distinguishing the most imporant radical inventions from the much larger number of incremental improvement inventions, it is also desirable to use a separate measure of key inventions (Baker, 1976 and Clark, Freeman and Soete, 1981). This was done at the NIESR with the assistance of a specialist consultant, Dr C. A. Redfarn. By this means 117 'major technical advances' were identified over the period 1790–1955. An 'advance' might be embodied in one key patent or in several related patents. Of these 117, IG Farben were responsible for thirty out of the German total of fifty-one, Du Pont for twelve out of the American total of forty-three, and ICI for seven out of the British total of fifteen. All other countries accounted for only eight.

Patents and innovations

The evidence of the patent statistics and key technical advances can now be compared with the achievements in 'innovation'. The simplest method of measuring 'innovations' is to list all the new synthetic materials and to identify that firm in each country which was responsible for the first commercial production. This gives a list of innovations but it is, of course, subject to the criticism that it omits most 'innovations' in new applications of a material, and in new processes of manufacture. Nevertheless, it provides a rough guide to innovative achievement, and can be adjusted to allow for the relative importance of each material.

Hufbauer prepared such a list of fifty-six plastics, synthetic rubbers and fibres and identified the first producer in a large number of countries. One interesting result of his study was that it enabled comparisons between countries and firms not only in terms of numbers of 'innovations' (world's first producer) but also in terms of 'imitation' (first producer in a particular country after the innovation). Hufbauer used the data to measure 'imitation lags' and demonstrated the extremely important result that countries with the highest 'innovation' rate also had the shortest 'imitation lag'. On the average it was only about three years before Germany or the US 'imitated' an innovation made abroad, but for Britain and France it was several years longer, and for all other countries more than ten years.

For most it was more than twenty. This result has major implications for the theory of foreign trade, which were explored by Hufbauer and other economists (see Hufbauer, 1966; Vernon, 1966; Posner, 1961). Size of national market is, of course, also an important factor affecting the imitation and diffusion process. Here we are concerned primarily with 'imitation' as an additional indicator of research and innovative capacity.

A firm with a strong research capacity may be able to assimilate and imitate more quickly. It may also innovate almost simultaneously. Thus, during the Second World War IG Farben were able to launch independently the production of poly-ethylene (ICI innovation) and of mylon (Du Pont innovation), while ICI was able to launch production of PVC in 1940 and the US chemical and rubber in-dustries were able to launch a range of synthetic rubbers (IG Farben innovations).

Comparing the patent statistics with those for 'innovation' and 'imitation', it is clear that IG Farben scores well on almost all counts (Table 3.5), and so to a lesser extent do Du Pont and ICI. The IG performance is signficantly better in numbers of innovations than in patents and much better in innovation than in imitation. The more successful a firm has been as an innovator in new products the less need for it to be an imitator, even where it has the capacity to imitate. The profit margin on original innovations may be much better than on imitation,[5] and the lead-time over competitors encourages the firm to concentrate on its own new products. However, this must be balanced against the high risks and heavy losses associated with unsuccessful original innovations, as in the case of Corfam discussed later in this chapter. Alternative strategies relating to imitation are dis-cussed in chapter 8.

Where a major chemical firm does imitate the innovation of a foreign competitor, whether under licensing arrangements or independently, it will often attempt to develop a better process or a major improvement in the product. In some cases the modification or improvement may be just as important as the original innova-tion. When cellophane was licensed to Du Pont in 1924, two of their research chemists found a way to make it moisture proof, which greatly increased its range of applications as a packaging material. The IG Nylon 6 differed in important respects from Du Pont's Nylon 66 and was the result of an independent R & D programme.

Furthermore, it must always be remembered that after the introduction of a new process or the construction of a new plant, many minor technical improve-ments will be made. These will not necessarily be recorded in patent statistics, both for reasons of secrecy and of patentability. Hollander found that in the case of Du Pont's rayon plants, many of the minor technical improvements were not patented but were together more important in their contribution to productivity than the major changes. Moreover, they were mainly initiated by the Engineering Department or Technical Assistance Groups, rather than Central Research. In the absence of any detailed data on the minor technical improvements in IG Farben plastics plants, it is not possible to assess IG's performance in minor innovations, except indirectly from their general competitive performance by comparison with the world industry (including exports) and by that part of minor innovation which is reflected in patent statistics.

[5] An ICI study of the productivity of its research found that the highest yield was from new products and processes based on in-house research. Although it is true, as Mueller has demon-strated, that many of Du Pont's innovations were not the result of its own R & D, nylon made by far the biggest contribution to corporate profits (see Holroyd, 1964; Mueller, 1962).

Table 3.5   Patents and innovations in synthetic materials (percentage world total)

Patents and innovations	Total	Percentage world total		
		IG Farben[a]	Du Pont	ICI
All plastics patents taken out by firms 1791–1945	6777	20	6	2
All plastics patents taken out by firms 1931–45	4341	20	8	2
'Major technical advances' in patent literature 1791–1945	117	26	10	6
Innovations in synthetic materials 1870–1945	56	32	9	2
'Major innovations' 1870–1945	20	45	10	5
Innovations 1925–45	36	44	11	3
First 'imitations' 1870–1945		14	4	8

[a] Including predecessors and successors.
Source: Author's estimates from Delorme (1962), Hufbauer's analysis of innovations (1966) and Redfarn survey (see p. 60).

The fact that IG scores much better in major innovations and in major technical advances than in patents, and probably better in innovations than in proportion to its R & D expenditure, may reflect in part its efficiency as an innovating organization, but it also probably reflects its dominant monopolistic position in the German war economy. Not only did IG dominate the various trade associations and consultative bodies for the chemical industry, it also controlled some important marketing outlets for its new products. This enabled it to exploit some of the cumulative advantages of scale in R & D programmes. With its immense resources and government backing for the synthetic substitute programme, IG was assured of a market for some of those synthetic materials which it successfully developed, especially synthetic rubbers.

This was by no means true in the US or UK, except during the Second World War itself. Even then the greater availability and lower price of many natural materials lent less urgency to the development of synthetics. However, the influence of a strong government-backed demand was also seen in the case of polyethylene in the UK and to a far greater extent in the synthetic rubber programme in the US, when Far Eastern natural rubber supplies were temporarily interrupted.

At one time in the 1920s IG had offered to sell its synthetic rubber patents to the world natural rubber cartel in the belief that the natural product would retain its major price advantage, but with large-scale rearmament, development work was pursued with extreme urgency and a major new industry was established, which was essential to the German war economy. However, it is notable that several of the most important plastics were developed by IG *before* Hitler's accession to power, including PVC, polystyrene, polyvinyl acetate, urea-formaldehyde glues and melamine formaldehyde. Consequently, IG's outstanding performance cannot be explained purely in terms of exceptional wartime demand, or rearmament, unless it is postulated that covert rearmament dominated IG research policy even in the 1920s.

Throughout the 1950s and 1960s West Germany continued to enjoy the world's highest level of *per capita* production and consumption of plastic materials, although her synthetic rubber production was temporarily interrupted after the war. The IG Farben Trust was dissolved but the successor firms continued to make important innovations and develop new applications. Thus the overall picture which emerges is one of an industry in which the innovative process was dominated since the First World War by the largest chemcial firms with strong professional R & D facilities. However, it would be misleading to think of the process in terms of massive planned research programmes leading almost automatically to the introduction of new materials. A high degree of technical and commercial uncertainty was characteristic of all these innovations, as is quite evident when we consider three of them in a little more detail.

PVC

A good example of IG Farben's research and innovation is the story of PVC, one of the three highest tonnage plastics in world consumption. It has been relatively neglected in the Anglo-American literature, but a study by a British chemist gives a valuable detailed account which is summarized here (Kaufman, 1969).

Vinyl chloride was first prepared and described by a young French chemist, Regnault, at Liebig's laboratory in 1835. Subsequently, the polymer was also described by other academic chemists but without any inkling of its potential industrial preparation or applications. In 1912 and 1913 an industrial research chemist, Fritz Klatte, working at Griesheim (Hoechst), took out a series of patents which anticipated the industrial processes used twenty years later and some of the future applications of PVC. However, PVC is an unstable material which deteriorates on exposure to light, is extremely hard to work and may liberate hydrochloric acid when heated. The monomer is difficult to prepare and Klatte's polymerization process was unsatisfactory. It was not until these problems were solved and suitable plasticizers, stabilizers and compounding elements were developed that PVC could find extensive commercial applications. It was mainly in the laboratories of IG Farben that solutions were found to these intractable problems in the 1920s and 1930s, and then only when fundamental knowledge of macromolecular structures permitted a higher degree of understanding and control of the process.

Laboratory work continued at Hoechst during the First World War, in the hope of developing substitutes for natural materials during the blockade. About four tons of polyvinyl chloracetate were produced at Griesheim. All three of the leading German chemical firms had research programmes on synthetic rubber well before this, and Bayer were successful in producing the first synthetic rubber, methyl butadiene, during the war. Production was discontinued in 1919, as natural rubber was still superior in quality and much cheaper. In the case of PVC, Klatte's patents were allowed to lapse in 1926, although research work continued. At this time other vinyl polymers appeared to be more promising and polyvinyl acetate was produced on a commercial basis by both Hoechst (IG) and Wacker in 1928-9. This success, together with the deeper understanding provided by Staudinger's papers on macromolecules, led to a renewed interest in PVC and co-polymers, and to the first commercial production early in the 1930s.

Early efforts to polymerize vinyl chloride had been based on thermal methods, on photopolymerization (as in Klatte's original patents), or on solution polymerization (as in the Russian chemist Ostromislensky's patent of 1912). Although further patents were taken out on all these methods during the 1920s, notably by Du Pont, the breakthrough to a commercial production process came with the development by IG Farben and Wacker of dispersion polymerization processes. The emulsion polymerization process developed by Fikentscher at the IG Ludwigshafen laboratory from 1929 to 1931, although now superseded, was used in the full-scale plant at Bitterfeld throughout the Second World War.

The successful development of this process owed a good deal to the earlier work on the rubber Buna-S, a butadiene-styrene co-polymer, at Bayer, as well as to the close cooperation between Staudinger and IG Farben's research workers. This was a case where the cumulative benefits of research experience on a variety of polymers and co-polymers (acrylonitrile, styrene, methyl methacrylate, vinyl acetate, vinyl chloride, etc.) greatly facilitated IG's progress. In a few areas other leading chemical firms, particularly Union Carbide, B. F. Goodrich and Du Pont in the United States, and ICI in Britain, were not far behind IG. Indeed, some of the early advances in PVC plasticizers were made at B. F. Goodrich rather than at IG Farben. But German research on plasticizers overtook Goodrich and was well ahead in the applications of PVC and co-polymers, not only for sheathing and insulation in the cable industry, but also in soles for footwear, in floor tiling, chemical plant, packaging and many other uses. An extremely important early application was in the development of magnetic tape as a result of a joint research programme with the German electrical firm AEG. All of this work, of course, enjoyed a powerful stimulus after 1933 from war preparation and fears of shortage of natural materials.

From this outline it is apparent that it is hardly possible to speak of the 'invention' of PVC in the same sense as the 'invention' of the safety razor. Its successful introduction into the German economy was the result of a long series of experiments, inventions and discoveries extending over a period of thirty years with many set-backs and disappointments. It was really not one but a family of materials and even during the Second World War their use was still on a relatively small scale, reaching only about 10,000 tons early in the war. Even after the war, the possibilities of PVC were often greatly underestimated, partly because of the very poor quality of some early products in such fields as rainwear. It was the end of the 1950s before world production reached the level of a million tons. In the postwar period many further improvements were made in the production process, notably as a result of the substitution of ethylene for acetylene as the basic intermediate and the use of suspension polymerization.

US, German and British firms shared in these process developments and in the applications research which led to the establishment of PVC as a major material in construction (drain-pipes, gutters, flooring, etc.), clothing and footwear, packaging and engineering, as well as wire and cable. It is now second only to polyethylene as a bulk tonnage plastic material.

Polyethylene

The most important single plastic material was not deliberately sought but was the indirect result of other research.

The links between chemical research and ICI's polyethylene innovation have been thoroughly discussed by Allen (1967). Like nylon, polyethylene also owed its discovery to a programme of fundamental research. But, unlike the Du Pont programme and Staudinger's research, this was not oriented to work on the structure and synthesis of long-chain molecules, but to the study of high pressure reactions. The Alkali Division of ICI had a strong research tradition from its origins as Brunner Mond. Mond himself had made a number of improvements in the Solvay ammonia-soda process and established the research laboratory at Winnington. The Research Director in the 1920s, Freeth, encouraged a long-term approach to research and in 1925 recruited a young British research chemist R. O. Gibson, who had collaborated closely with Professor Michels of Amsterdam. When Gibson began to work at ICI this association continued and Michels became a consultant for ICI, because of his outstanding work on the effects of high pressure. Special equipment designed by Michels was installed at the ICI laboratory at Winnington in 1931 and was used in a series of studies of chemical reactions.

In the course of these experiments, polyethylene was discovered and its properties recognized. The discovery owed a good deal to chance in that on one occasion in 1933 a defect in the apparatus led to the polymerization of ethylene. The research engineer, W. R. D. Manning, made important improvements in the apparatus, making it relatively safe to continue the high pressure experiments. It would be impossible to identify any one man as 'the inventor' of polyethylene, as Swallow, Fawcett and Perrin also made important contributions to the work. Patents were taken out in 1935, and the difficult and expensive development of a safe high-pressure process took several years longer. Michels assisted in the compressor design which was critically important for the pilot plant which started production in March 1938. Although ICI were quick to identify some important properties of the new material, it was a long time before most of the present-day applications of polyethylene were realized or even conceived. Although its possibilities as an insulator were recognized almost immediately, it was expensive to make and before the war it was assumed that its main application would be in submarine cables. Joint applications development with BICC had established its importance in this field already by 1937. It proved extremely important in wartime radar applications, but with the post-war drop in defence demand, the closing down of the main ICI plant was apparently seriously considered.

However, the versatility of the material was gradually recognized and many new applications were established in the 1950s as process development work lowered the costs of production, and applied research led to the necessary modifications and new grades of the basic polymer. A powerful impulse to the growth of the world market also came from the decision of the United States courts in 1952 to compel ICI to license several other US chemical firms, in addition to Du Pont, the original licensees. Although bitterly contested at the time, on the grounds that the court had no jurisdiction over ICI, the decision may well have been a blessing in disguise even for ICI, in that it almost certainly led to a more rapid growth of new applications, particularly in the domestic field and in packaging, as well as to a substantial increase in licensing fees and know-how payments. Among those who took licences were Monsanto, Dow, Koppers and Spencer.

A further stimulus to the growth of the total world market for polyethylene came from the introduction of many process improvements. These affected both the cost of production and the range and quality of new applications. Both Union Carbide and Du Pont, after originally obtaining licences from ICI, changed the technique to such an extent that they could then themselves grant licences on the basis of their own technology. As early as 1938, BASF had an independently

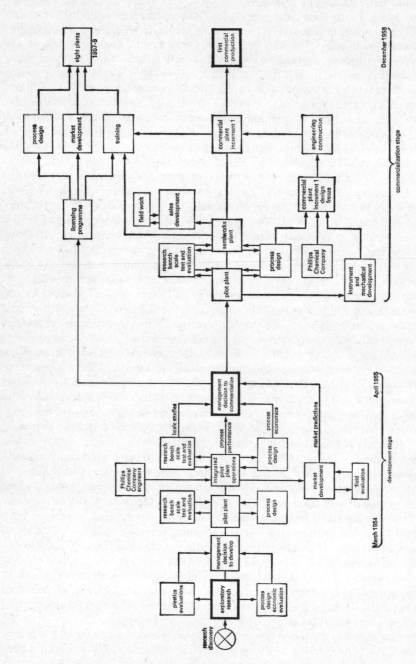

Fig. 3.1   Project flow sheet: medium pressure polyethylene. *Source:* Phillips Petroleum Company.

developed pilot plant using a somewhat different type of reactor (Schott and Müller, 1975). Many new types of polyethylene were introduced to suit the special requirements of particular end-uses and these depended on the ability of the resin producer to control the melt index and density very closely. Product and process improvements were thus intimately related. Schott and Müller (1975) distinguished as many as ten distinct high-pressure processes, but in the 1950s came the more radical innovation of low-pressure (high-density) polyethylene and the Phillips Petroleum medium-pressure process. Although there were some fears at the time that the original ICI high-pressure process might be adversely affected by these new developments, it turned out that they were partly complementary rather than competitive. The new types of polyethylene had a higher density than the original ICI low-density 'polythene' and rather different properties. Thus they tended to enlarge the total range of applications rather than to diminish the importance of the high pressure process. Although it was never as important as the major processes, interesting aspects of the Phillips development were the decision to license simultaneously with the first commercial production (Figure 3.1) and the speed of development.

As a result of a continuing programme of intensive R & D by ICI and its many licensees, numerous improvements in both products and processes continued to be made. Although the original patents had long since expired (including the special extension which was granted because of the war), ICI continued to enjoy a considerable income from the sale of know-how and the use of more recent patents throughout the 1960s. The technical leadership of the company and the continued importance of minor improvements were such that new producers still found it desirable to make substantial payments for this know-how. Although the total amount of these payments has not been revealed, together with those for Terylene they must have accounted for a substantial proportion of the £10 million or so which ICI was receiving for licence and know-how payments annually in the 1960s. (In 1971 total ICI receipts from licences amounted to £13 million and expenditures for licences were £3 million (ICI, 1971).) During the 1970s the situation in high-pressure polyethylene underwent a drastic change. The pressure of rising costs and the entry of many new producers led to the erosion of profit margins and substantial over-capacity. Britain was overtaken and outstripped by other industrial countries. The newer processes proved more profitable.

Ziegler's low-pressure process was discovered in 1950 in the course of a programme of research on catalysts at the Max Planck Research Institute in Mulheim. He was at the time already a research consultant with Hoechst and arrangements were soon made for large-scale development of the new process based on the use of his aluminium catalyst and his patents. Production began in both Germany and the US in 1956. The combined use of all types of polyethylene now makes it the largest tonnage plastic in the world, accounting for over 10 million tons by 1980.

Further work on catalytic polymerization resulting from Ziegler's discoveries led to the development of a process for the synthesis of polypropylene by Dr Natta of Milan Polytechnic, working closely with the leading Italian chemical company, Montecatini. This process, too, was widely licensed throughout the world, but some US companies contested the patents and claimed that they had developed a polypropylene process independently. Its rate of growth was extremely rapid, as it is cheap and tough and provides worthwhile applications for a refinery by-product (propylene) which was often wasted before. It soon became one of the major bulk tonnage plastic materials along with PVC, polyethylene and polystyrene.

Corfam (synthetic leather)

This last example differs from the others in that it turned out to be a commercial failure, and resulted in a loss estimated at $100 million. It is all the more instructive for this reason. Encouraged by their outstanding success with nylon, by the late 1930s the Du Pont Central Research Department had developed several techniques to make porous poromeric films, and work continued on these techniques throughout the 1940s (Lawson, Lynch and Richards, 1965). Permeable films were wanted not only for leather substitutes for shoe uppers but also for other textile, packaging and coating applications. Several of Du Pont's product groups (industrial departments) collaborated with the Central Research Department in development work in the early 1950s, but from 1956 the work was concentrated in the Fabrics and Finishes Department. Field trials in 1956 and 1957 for applications in shoes and garments led to a decision to go ahead with full-scale development and pilot plant construction in 1959. Big engineering problems had to be resolved in scaling up production and it took longer than anticipated to achieve a satisfactory process.

Leather prices were rising and market studies suggested that the potential market for grain leather-substitute shoes was large and stable. A sales manager was appointed in 1960 and regular quarterly sales–research product planning sessions were held to iron out the difficulties involved in preparing product launch. Initially it was decided to concentrate on the higher-priced women's fashion shoes. 'We felt this marketing concentration would create a much more desirable effect than just having shoes shipped helter-skelter into all markets.' An intensive study was made of the economics of the shoe industry and of the leather market, and another special study on the attitude of consumers to the retail price of shoes.

More than 16,000 pairs of shoes were made in 200 different shoe factories and extensive field trials carried out to ensure that the product was acceptable, hygienic and comfortable. A computer 'venture analysis' model was used to test out various assumptions about future market growth, the effects of advertising and other factors. Finally, after a tremendous advertising campaign full-scale production was launched at a plant in Tennessee, and sales built up rapidly both in North America and Europe in the period 1966 to 1968.

However, although several million pairs of shoes were sold using Corfam, the product did not bring the rewards anticipated by Du Pont and in 1970 it was announced that the plant would be closed down and the company would withdraw from the market. No full explanation has ever been given but it may be conjectured that the availability of cheaper synthetic substitutes (PVC and Porvair) for ladies' shoe uppers was one of the factors leading to this decision. Porvair itself, although temporarily rather more successful than Corfam, continued to face great uncertainties and risks throughout the 1970s (Gibbons and Littler, 1979).

The example is instructive mainly because it demonstrates that neither strong R & D, nor experience in innovation, nor thorough market research and trials, nor careful new product planning can in themselves ensure success in innovation. The implications of this degree of uncertainty confronting even the largest and most successful innovating firms are discussed in chapter 7.

Product development and process development

This summary account of the background to the innovation of several of the major synthetic materials has concentrated on the materials themselves and their applications. But it has become evident that their successful development, and perhaps even more the shift to very large tonnage plants, depended upon the kind of innovative process engineering described in chapter 2. New processes had to be developed not only for the new materials, but also for the intermediate products used in their manufacture. To a large extent it was the chemical firms themselves which also undertook this engineering work, although they collaborated with other firms for particular items of process equipment such as compressors, pressure vessels and valves. It involved the specialization of the plant design and engineering functions and the consequent emergence of the new profession of chemical engineering.

During the 1960s and 1970s, the evidence of patent statistics suggests that R & D efforts were increasingly concentrated on process improvements rather than the search for additional new materials (Walsh, Townsend, Achilladelis and Freeman, 1979).

The new materials themselves were often first isolated or discovered by relatively inexpensive methods, or indeed even accidentally. Frequently this occurred in university or other laboratories concerned with fundamental research, often long before commercial application. The reason for the proportionately large contribution of the giant chemical firms to the *innovation* of these materials on an *industrial* scale lay in the expense and difficulty of developing a satisfactory *process*, both for the material and for the intermediates and by-products. In every one of the cases which we have considered this proved to be very costly and it took a long time. It required a combination of skills in fundamental physical chemistry, process design, mechanical engineering and other types of engineering. It was difficult to bring together such a combination of skills except within the framework of a professionalized R & D system in industry (Figure 3.1), or within the framework of a public corporation such as the Atomic Energy Authority.

Instruments

The increasingly intimate relationship between new materials, new process development and fundamental research is nowhere more apparent than in the field of instrumentation. It would have been impossible to develop nuclear power or many new chemical processes and materials in the last fifty years without new scientific instruments. The use of on-line chromatography, analogue controllers, and specialized sensors and transducers has revolutionized chemical plant design and made possible the precise monitoring and remote control of complex flow processes. It has also led to enormous improvements in standards of purity and in quality control.

The design of new laboratory instruments, particularly spectroscopes, was an integral part of the advance of physical chemistry. The development of mass spectrometers, other types of spectrometer and the electron microscope made it possible to ascertain molecular structure and the arrangement of specific groups within a molecule. The innovation of these and other new types of instrument strengthened the links between chemical technology and fundamental scientific knowledge, because most of the instruments originated in basic research laboratories. As Shimshoni (1970, p. 64) has pointed out:

Physical chemical analysis began with the observation of the visual part of the spectrum. It was found very early that substances could be identified by the characteristic special patterns of the wavelengths emitted when a sample of a substance is excited, or of the wavelengths absorbed from an external source by a sample. The subsequent story of the applications of physical methods to the study of matter is that of the extension of the spectral range until the whole of the electromagnetic spectrum could be used, of obtaining increased efficiency in discrimination or detection, and of the design and marketing of lower-cost instruments.

In the course of their fundamental chemical work, the universities frequently designed and built new instruments for their own use, and these were often commercialized by scientist–entrepreneurs setting up as instrument manufacturers. Many of the important instrument companies in the United States and Europe started in this way.[6] Sometimes scientists from chemical companies also assisted instrument companies in the development of new products. The particularly intimate nature of this collaboration has been documented in detail by von Hippel (1976 and 1978). Together with university laboratories and government laboratories, the R & D departments of chemical and oil companies were the main *users* of laboratory analytical instruments. The infra-red spectrophotometer, which was innovated by Perkin–Elmer in 1943, was largely designed by Barnes and Williams of Cyanamid Laboratories, who needed the instrument for their analytical work. The role of scientists who set up their own instrument companies was decisive and Daniel Shimshoni has documented in detail the critical role of these inventor–entrepreneurs in establishing the main products.

Most of the recent advances in laboratory nuclear and process instrumentation have been in the development of *electronic* instruments, and the electronic computer is now of critical importance in laboratory and design calculations, as well as in process control. This is the link between the industries which we have so far discussed and the electronics industry to which we now turn in the next chapter.

We may note that the huge increases in the price of oil during the 1970s and other induced changes in the relative prices of energy and materials have provided a major new stimulus to the R & D of the chemical industry. This new concentration of effort on energy-saving technologies has already led to considerable economies in the use of energy and has affected the introduction of major new processes, such as ICI's new chlorine cell, the FM 21. In the closing years of the twentieth century or in the twenty-first century, it will in all probability lead to a resurgence of processes now under development, using coal as a feedstock once more.

[6] For example, Perkin–Elmer and Hewlett–Packard.

4 Electronics

Electronic communications and control systems make it possible to perform in minutes or even in a fraction of a second calculations and operations which previously took weeks, months or years, or could not be performed at all, and to perform them with a higher degree of reliability and at a lower cost than by older methods (Table 4.1). Beginning with radio communications in the 1890s and television in the 1930s, the applications of electronics have spread first to systems of detection and navigation (radar), and since the war to computers for data processing and to the control of a great variety of industrial processes.

Table 4.1 Technical progress in computers

Measure	Vacuum-tube computers (early 1950s)	Hybrid integrated circuits–360 system (late 1960s)
components per cubic foot	2,000	30,000
multiplication per second[a]	2,500	375,000
cost ($) of 100,000 computations	1.38	0.04

[a] A single multiplication on mechanical or the first electromechanical computers took more than one second (see p. 81).
Source: Fortune, September, (1966).

The introduction of reliable low-cost electronic computers into the economy is the most revolutionary technical innovation of the twentieth century. Whilst it is true that the older mechanical and electro-mechanical calculators and other devices could already perform some of the functions of modern computers before and during the second world war, it was the *electronic* computer which totally transformed both the range of potential applications and their cost. Table 4.1 illustrates the dramatic increase in capacity and reduction in costs of computing which occurred in the 1950s and early 1960s, from the early valve (tube) computers to those using semi-conductor technology and integrated circuits. Since that time the micro-processor revolution of the 1970s and 1980s has further increased the number of components per cubic inch and reduced their cost by at least two more orders of magnitude, so that the computers of the 1950s now appear incredibly expensive and cumbersome. Nevertheless, already in the 1960s electronic computers had greatly increased the efficiency with which enormous quantities of data could be stored and processed, such as payroll calculations, invoicing, insurance premiums, design calculations and so forth. In addition, electronic equipment is so fast and reliable that automatic 'feed-back' systems can be used to control operations in 'real time', even where this involves fairly complex calculations with several variables, as in chemical processes or aircraft navigation. A computer of some sort is often at the heart of such systems, and it is electronics which made it possible to 'automate' a much greater variety of operations and processes than was hitherto possible. In a sense, there is nothing really new in 'automation', as the thermostat, invented in 1625, already represented an automatic feed-back control system. It is no accident that some of the pioneering firms in electronic equipment for automation also produced thermostats and regulators (for example, Honeywell and Elliott-Automation). The difference is one of degree. However, electronics has increased the applications of the automatic feed-back control principle so rapidly that it is not unreasonable to look upon 'automation' primarily as a post-war change, associated with the electronic computer, electronic sensing and detecting devices, and process instrumentation.

Experience in war-time radar, gun-control systems and missile guidance devices formed the basis for revolutionary advances in industrial process control-systems and 'numerical control' of machine tools. Together with new types of electronic instruments and robots these are gradually transforming engineering processes so that many of them will increasingly come to resemble automated flow processes. Thus, mechanical engineering, too, will be increasingly science-based and the old craft skills will cease to dominate the design and manufacturing system in the traditional 'heartland' of capital goods manufacture. The Molins 'System 24' designed by Williamson in the 1960s was an early forerunner of that complete automation of machine shops which is now being realized by a growing number of Japanese firms, based on a combination of computers, NC and robotics.

Since the Second World War, military applications of radar have been extended to complex early warning systems, missile guidance and so forth. Civil applications have also grown rapidly—in air traffic control, airborne and marine radar and navigation systems, space exploration and aids for fishing vessels (sonar). Both in civil and military applications there is also a close link between the growth of these installations and of computer networks. The American 'SAGE' (Semi-Automatic Ground Environment Control) system linked a big chain of radar stations to very powerful computers, and the same principle was used in 'NADGE' and in the civil air traffic control systems in Western Europe. On a smaller scale, airborne navigation systems such as the Decca 'Navigator' and Marconi 'Doppler' equipment make use of very small specialized computers.

Originally, the radio industry formed the basis of the electronics industry followed in the 1930s by television. Communication equipment and entertainment and information systems remain an extremely important part of the industry, but they too are being transformed by the electronic computer. The telephone system increasingly uses electronic exchanges and switching computers and is increasingly used for data transmission and for communication between computers. Services such as 'Prestel', despite their teething troubles, have already demonstrated the way in which domestic television and VCRs can be linked up through a greatly expanded telephone network to provide a wide range of information services both for households and for business. 'Tele-shopping' and 'tele-banking' are now clearly on the horizon and so is the mainly electronic office. Word processors are already in use on a large scale and electronic mail services are increasingly important for business communication. Consequently, electronic innovations in the twentieth century are revolutionizing service activities and households, as well as all branches of manufacturing.

It is seldom possible to talk meaningfully about the 'inventor' of any of the major electronic products. Their successful realization, first in a laboratory and later on a commercial scale, depended on contributions from many scientists and engineers in several countries over a long period. They are 'systems' involving a large number of components, all of which are subject to change and improvement. The interplay of component innovations, materials innovations, and new capital goods and consumer products is one of the most important features of the industry's development. In particular the introduction of transistors in the

1950s, and later of integrated circuits and micro-processors, facilitated major new advances in the design of electronic products, and reductions in their cost and size.

Any analysis of this industry must study innovation problems particularly carefully. For none of the products existed before the beginning of the century, and most of them did not exist in their present form even fifteen years ago. The entire industry is based on research and it is one of the most research-intensive of all industries (Figure 1.2 and Table 1.2). Like plastics, it illustrates the transition from the inventor–entrepreneur to the corporate R & D department. In the historical account which follows compare, for example, innovation in valves with semiconductors or radio with television. Along with the growth of corporate professionalized R & D, governments too have played a very big part both as sponsors and directly in the conduct of R & D. This has culminated in the space research programmes, but was already very important in radar. However, the inventor–entrepreneur continued to play a significant role in industry, especially in electronic instruments. Even in computers his role has been more important than is commonly assumed. The account below briefly summarizes the major developments in each of the five main sectors—radio, television, radar, computers and components—and concludes with an analytical discussion of the economies of scale in research and innovation in the industry.

Radio[1]

For radio, Maxwell's theory of electromagnetism in the 1860s was the foundation for Hertz's first practical laboratory demonstrations of the production and detection of wireless waves in Germany in the 1880s, for Branly's coherer demonstrated in 1890, and for Lodge's demonstration of wireless reception at the British Association in 1894. At about the same time Popoff demonstrated an improved system of reception at the University of Kronstadt. All these men were academic scientists engaged in fundamental research. It was not until Marconi formed his Wireless Telegraph Company in London in 1897 that systematic applied R & D work began. It was Marconi who gave the first practical demonstrations of wireless communication between ships and shore, between shore stations, between countries and finally across the Atlantic in 1901. It was his new company which established the first regular wireless telegraph services, both between countries and from ships, followed closely by Telefunken in Germany. His role in the electronics industry was analogous to that of Baekeland in plastics, as a highly successful inventor–entrepreneur.

Up to and during the First World War, there were innumerable improvements in the components, circuits and techniques used in radio communication, made by inventors in many countries, but the most important of them originated in Britain, Germany and the United States. It would be difficult, if not impossible, to evaluate precisely the relative contribution from scientists and engineers of the three principal countries. Some of the developments, such as the feed-back circuit, were almost simultaneous in all three countries, and there was bitter patent litigation which went on for twenty years. While Professor Fleming of University College, London (who was employed as a consultant to the Marconi Company), invented and patented the first thermionic valve in 1904, it was the American, de Forest,

[1] For the early history of radio and television see Sturmey (1958), Maclaurin (1949), Briggs (1961), Telefunken (1928, 1953), Radio Corporation of America (1963).

who invented the triode valve in 1907, which later proved far more effective for reception and transmission. Other American inventors, notably Langmuir, Armstrong, Alexanderson and Fessenden made important contributions to the improvements of valves, circuits, alternators, and aerials and in the introduction of radio telephony. The American Telephone and Telegraph Company (AT&T) pioneered the use of valves for relays in the telephone system, having purchased rights to de Forest's triode patents.

But in this early period, up to the end of the First World War, the leading companies in the industry were not American but British and German. The largest manufacturer of radio in the US was a Marconi subsidiary, and the world market was dominated by Marconi and Telefunken, who between them controlled most of the ship and shore installations all over the world, including those in the United States. In 1915 the Marconi companies controlled 225 out of 706 coast installations, and 1894 out of 4846 marine installations. Until the outbreak of the First World War the share of Telefunken was probably somewhat larger. Although starting a year or two later than Marconi, the largest German electrical companies, Siemens and AEG, rapidly developed effective radio communication systems— Professor Braun's system at Siemens and the Slaby-Arco system at AEG. They received strong backing from the German Government, particularly the Navy, and were persuaded by the Kaiser to merge their interests in 1903 and form the jointly-owned Telefunken Company. The company was concerned primarily with R & D and with the sale, installation and maintenance of radio stations. AEG and Siemens continued to manufacture parts and equipment. Already by 1906 Telefunken had more installations than Marconi. While the Marconi patents were very strong and upheld in most countries, their priority was not accepted in Germany until 1912 when Telefunken and Marconi reached a world-wide agreement on patents, licences and know-how. Similar cross-licensing and know-how agreements were renewed after the First World War in 1919, and again more recently.

Thus both Marconi and Telefunken had well-organized industrial R & D programmes and had strong patent positions. They were able to assimilate and imitate the technical advances made in other countries. Both of them were able to provide world-wide technical service, had their own schools for training radio operators, and were able to repair and maintain leased equipment. (The first Marconi school was opened at Frinton in 1901.) But Telefunken had the advantage over Marconi in consistency of government support and of its financial resources. Although the Post Office had originally encouraged Marconi in his first experiments in Britain, and the Admiralty had also been sympathetic, later on relationships with the Post Office became difficult and the attitude of the government was sometimes unhelpful and even obstructive. There were, of course, difficult issues involving the problem of private monopoly in communications and the Marconi company's determination to uphold and exploit its strong patent position.

Like many major innovations, radio did not prove profitable for a long time and the Marconi Company paid no dividends from its inception in 1897 until 1910.

In view of the recent tendency to ascribe almost superhuman attributes to US management, it is important to note that, by contrast with Marconi and Telefunken, the early American radio companies were poorly managed, and some of the pioneering inventors, such as de Forest and Fessenden, were failures as innovators and entrepreneurs. De Forest produced a stream of inventions and patents but, although his Wireless Telegraph Company had orders from the Navy and War

Department, it failed to produce reliable communications equipment. It was not until the formation of the Radio Corporation of America in 1919 that a really successful specialist electronics enterprise was started. The big electrical companies, General Electric and Westinghouse, and the major telephone company, American Telephone and Telegraph, all had an interest in radio and relatively strong R & D organizations which had made major contributions to the development of radio telephony. They were blocked, however, by the fact that in international communications the British Marconi Company dominated the field, and in the United States the control of key patents by opposing interests contributed to a stalemate that retarded the best utilization of radio. An imaginative solution to this deadlock was found as a result of government initiative. With strong encouragement from the United States Navy, Owen D. Young of General Electric set about buying out Marconi's American subsidiary and setting up a powerful unified American-owned radio company. The Navy's motivation was partly commercial and partly strategic. Both the Secretary of the Navy (who favoured a publicly owned communications network) and the Assistant Secretary, Franklin Roosevelt, wanted an American company because they could not accept a position where a vital communications network was controlled by a foreign, even though friendly, power. They and other officials of the Navy also realized the great commercial potential of radio.[2] Owen Young became chairman of the new Radio Corporation of America (RCA) with General Electric supplying sets and valves (Westinghouse came into the arrangements for manufacturing in 1921). David Sarnoff, of the old Marconi Company, became commercial manager (and later President), and RCA immediately concluded patent and cross-licensing agreements with British Marconi, Telefunken and the French Compagnie Générale de Télégraphie sans Fil (CSF), which was also based on a former Marconi subsidiary. After some bargaining, agreements were also reached within the United States that ensured that RCA had the right to use over two thousand American patents, including all the important ones; RCA also made a cross-licensing arrangement with AT&T.

Shortly after the formation of RCA, the radio industry was transformed by the growth of public broadcasting. The United States industry took the lead in the scale of manufacture and improvement of design of home radio receivers, although European countries were not far behind. American companies never achieved that supremacy in the world export market for consumer goods which they later achieved in capital goods. Many of the important inventions concerning radio receivers between the wars were made by European companies, especially Phillips and its subsidiaries. In 1937 Dutch radio exports were as big as those of the US. After the Second World War, Japanese companies were quick to appreciate the possibilities of using transistors in radio sets, taking American licences for the semi-conductor devices, and achieving a very large share of world exports of electronic consumer goods.

The big post-war developments in capital goods were the introduction of worldwide short-wave communication networks in the 1920s and the introduction of frequency modulation (FM) in the 1930s. Short-wave communications were

[2] Owen Young's account states: 'When Admiral Bullard arrived in my office, he said that the President, whom he had just seen in Paris, was concerned about the post-war international position of the United States and had concluded that three of the key areas on which international influence would be based were shipping, petroleum and radio. But in radio the British were now dominant and the United States, with her technical proficiency, had an opportunity to achieve at least a position of equality.' (Maclaurin, 1949, p. 101.)

developed primarily by amateurs and by the Marconi Company. The development of directional aerials by Franklin at Marconi's made it possible for them to propose a Commonwealth radio chain using a beamed short-wave system. His proposals were accepted by the first Labour Government in 1924 in spite of opposition from the Post Office, which was wedded to cable communications, and in 1926 the first stations were opened. One of the immediate effects was a sharp reduction in cable rates; and eventually the Post Office took over the entire 'Imperial chain'.

Frequency modulation was pioneered by an independent inventor, Edwin Armstrong, who was a professor at Columbia University. He did not meet with much sympathy from the broadcasting networks or from RCA, and consequently had to build his own station to demonstrate his system. Partly because of these delays the first experimental operational FM network was set up not in America but by Telefunken for the German Army in 1936. Its success led to the establishment of a large-scale FM network during the war which covered the whole of German-occupied Europe and Africa. The introduction of FM in Britain came very much later in 1955.

Development in radio communication equipment since the war has been mainly in improved performance of existing systems rather than in completely new devices. The introduction of new components has, however, meant a continuing high rate of technical change and communication satellites are leading to further big changes.

Television

The possibilities of television were foreseen as early as 1884, when Paul Nipkow took out a patent in Berlin for his invention of the scanning disc. The invention and improvement of the photoelectric cell, and of the cathode-ray oscilloscope, also took place mainly in Germany, but it was Professor Boris Rosing of St Petersburg who first suggested using Braun's cathode-ray tube to receive images from a remote source, in 1907. Similar suggestions were made by Campbell Swinton in England. Zworykin was a pupil of Rosing's, who had already worked on a cathode-ray receiver before going to America in 1919. He had already conceived a complete electronic system for transmission and reception and in America patented the iconoscope which made it possible to transmit television pictures successfully. But it was not until he began a much more ambitious R & D programme at RCA that the numerous development problems were overcome, in the years from 1924 to 1939. RCA launched television commercially in 1939, but pictures had been successfully transmitted on a laboratory basis many years earlier.

Although EMI in England started later, they moved more quickly and a team led by Blumlein independently developed an iconoscope known as 'Emitron'. Marconi cooperated in the development of transmitters and the two firms were so successful in developing their system that the BBC were able to begin regular television broadcasts in 1936.[3] In the same year EMI and RCA made a licensing and know-how agreement. There were already financial links, and Sarnoff was a director of EMI. Telefunken, too, were able to develop an electronic television system before the Second World War and had a licensing and know-how agreement with EMI. Experimental transmissions were made of the Olympic Games in 1936 and

[3] Broadcasts using the less satisfactory Baird mechanical system had begun as early as 1929, and there was a ridiculous xenophobic campaign in the 1930s to try and persuade the BBC to discriminate in favour of the 'British' mechanical system against the allegedly 'American' EMI System.

regular broadcasts were made to some German troops in 1939–40. The war prevented the introduction of a regular service, although Telefunken had made advanced preparations for the mass sale of a popular model. Thus, although Zworykin's contribution to television was outstanding and other American inventors, such as Farnsworth, also made important advances, European countries were close behind America, and a public service was launched earlier in Britain than in the United States. It was in the 1939–41 period and the early post-war years that the American industry went ahead.

The part played by RCA in the development of television is particularly notable. Between 1930 and 1939 RCA spent $2.7 million on television research and development, and a further $2 million on patents and patenting. Another $1.5 million was spent on testing the system. Telefunken and EMI probably spent similar sums in the pre-war period. Such an investment was impossible for most smaller firms or independent inventors, but Farnsworth is said to have spent $1 million privately. The teams at RCA and EMI were quite small in the early days—Zworykin had only four or five assistants before 1930, and Shonberg, the research director at EMI, had only a few when he started in 1931. But the scale of effort built up as the introduction of a commercial system approached; the problems could not have been solved without resources of a fairly large organization.

This applied even more to colour television. RCA spent $130 million on launching colour television before it became profitable in 1960. After about four years' development work they demonstrated their first colour tube in 1950, but the Federal Communication Commission (FCC), after considering two competing systems, mechanical non-compatible and all-electronic compatible, gave its approval to the first, thus in effect banning the RCA system from the market. This decision was only reversed after several years, through court action by RCA. During this period, development work continued intensively and the first sales were made in 1954. Growth was very slow at first, because of the attitude of the broadcasting networks and the high price of sets, but sales had reached about five million sets per annum by 1970 in America.

Europe and Japan were both originally a good many years behind the United States with colour television, but Japan not only closed the gap in the 1960s and 1970s but went ahead to undisputed world leadership in this industry. By 1977 Japan accounted for over half of world production in colour television and three-quarters of world exports. They were exporting about 5 million sets, compared with about 1 million from Germany and 250,000 from the UK, despite the fact that Japan was still limited in many markets in Europe by the PAL patents and other restrictions. Later in the 1970s, Japanese direct exports declined especially to the US, because of Japanese investment overseas and because of agreements betweeen Japanese and American producers.

As in the case of the introduction of transistors into the radio industry, this extraordinary Japanese success was not based on simple carbon copy imitation, but involved a whole series of product improvements and process innovations. After comparing the performance of the American, European and Japanese industries, Sciberras (1980) concluded:

Japanese firms have been the most successful innovators [in the 1970s]. By applying advanced automation in assembly, testing and handling to large production volumes, the Japanese have achieved drastically superior performance in terms both of productivity and of quality.

Although he found that the main advantage of the new automated techniques was in improving product reliability, Sciberras calculated that Japanese man-hours per set were 1.9 compared with 3.9 in Germany and 6.1 in the UK. He attributed the opening up of this remarkable productivity gap mainly to the integrated approach to automation technology and to the intensive training of personnel at all levels in Japanese firms. Peck and Wilson (1982) also point out that the Japanese manufacturers were the first to introduce integrated circuit technology into the colour television industry (with the important economies in assembly labour that this involved). The success of this innovation was based on a joint research effort starting in 1966 and involving five television manufacturers, seven semi-conductor manufacturers, four universities and two research institutes and the overall backing of MITI (the ministry for trade and industry). This example illustrates the capacity of the Japanese social system to achieve a flexible mobilization of resources to make and diffuse decisive innovations quickly (Allen, 1981.).

Radar[4]

As with television, so with radar: the possibilities were conceived long before they were realized in practice. As early as 1904 a Düsseldorf engineer, Christian Hülsmeyer, took out a patent for a process for 'detecting distant metallic objects by electrical waves'. But he failed to produce a working prototype. In a lecture in June 1922, Marconi also foresaw the use of such a system for detection in darkness or in fog, but the Marconi Company did not do any development work in this field before the Second World War. In 1923 an American engineer, Loewy, also patented a radar device in America.

It was only when a government-sponsored R & D programme was started in Britain and in Germany in the 1930s that practical results were achieved. Radar (radio detection and ranging) became an invaluable military aid in the Second World War. Sir Robert Watson Watt demonstrated that the reflection from electromagnetic waves could be projected on a fluorescent screen (an oscilloscope), and received high level support for a crash programme to set up a chain of radar stations for air defence. British firms with experience in the development and production of high-powered transmitters (Metro-Vickers) and of cathode-ray tubes (EMI and Cossor) were associated with the programme before the war, while in Germany almost the entire R & D effort was concentrated in Telefunken. By the end of the war, eight to ten thousand people were engaged on R & D at Telefunken, but this included work on radio communications and control systems as well as on radar. Total numbers engaged were smaller in Britian and were mainly in government establishments, especially the Telecommunications Research Establishment (TRE), later the Royal Radar Establishment (RRE). Perhaps three thousand people were engaged in government establishments and another one thousand on radar development work in industry at the peak of the effort. Marconi's R & D staff were engaged entirely on communications work during the war. Because industry's development facilities were inadequate, the government teams often carried projects right through to production drawings, and sometimes undertook the first stages of manufacture as well. An extremely important feature of the whole programme was the intimacy of the direct contact between users of the equipment, manufacturers, and government (TRE) research teams, as embodied in the 'Sunday Soviets'.

[4] For the early history of radar see Postan, Hay and Scott (1964), Telefunken (1928, 1953) and Gartmann (1959).

The high priority assigned to radar work, the direct involvement of the best scientists in the country from universities and especially of Cockcroft and his colleagues from Cambridge, the relative freedom of the development teams and the large resources committed, resulted in an extraordinary rapid and successful flow of new devices and equipment, without previous parallel in the history of the industry. Among the new equipment successfully put into active service within a few years were the home chain of radar stations for intercepting enemy aircraft (CH), air interception equipment for fighter aircraft (AI), air-to-surface equipment for locating ships and surfaced submarines (ASV), equipment in ships and aircraft for identification as friend or foe (IFF), gun-laying equipment for control of anti-aircraft fire (GL), navigational and positional aids for aircraft and ships ('Gee' and 'Oboe') and display systems such as the Plan Position Indicator (PPI) which, combined with airborne radar, provided an accurate and detailed map of the ground below (H_2S, Home Sweet Home). Before his tragic death in an air crash, Blumlein led the development work at EMI on H_2S which was the most sophisticated type of radar.

Perhaps the most outstanding British achievement was the invention of the resonant cavity magnetron by a team at Birmingham University led by Professor Oliphant in 1940. This made possible the 'centimetric revolution', the use of very short wave equipment as low as 3 centimetres, which permitted far more accurate performance for various types of radar using smaller aerials. It involved an increase in the peak-pulse power of the radar equipment of several thousand times and was far more difficult to jam. This invention gave British radar a decisive lead over German equipment for a year or two, until captured equipment at Rotterdam permitted them to catch up. This invention also proved decisive in persuading the United States to pool all radar know-how during the war.

Although British radar developments were, in most cases, a little ahead of American and German work, the lag was short. All the main countries had been doing government-sponsored work in the 1930s. France had developed the first civil application of marine radar on the 'Normandie', and some radar know-how was brought to England in 1940. Germany had a number of operational radar devices at the outbreak of war, but they were relatively unsophisticated.[5] Tele-funken took out a patent for PPI already in 1936, but the work was not pursued energetically because U-boat chasers and night bombers were given a low priority in the German weapons programme. Work on centimetric equipment was also slowed down in 1940, as it was thought that it would be unnecessary. Once the centimetric revolution was taken seriously, with the aid of the Rotterdam equipment Telefunken soon caught up. The official British history records:

At the end of the war the Germans were developing a centimetric ground equipment of very high discrimination, against which no economic method of jamming could be foreseen. With this development the radar defence had caught up with the radar attack and the tactics of night bombing would, if the war had continued, have required drastic revision (Postan, Hay and Scott, 1964).

[5] For example, the Lorenz blind approach beam system used to guide bombers and its successor, X-Gerät, both needed continuous wave systems rather than pulse generation. These were less accurate and more easily jammed. Pulse techniques were not used until 1944. For similar reasons the German ground chain, Freya, using Würzburg sets (53 centimetres) was virtually immobilized by British bombers in 1943.

Table 4.2   Patents taken out in London by leading firms for radio navigation, radio-location and aerials

1947–50	No.	1951–4	No.	1955–8	No.	1959–62	No.
1 STC	112	1 Marconi Wireless Tel.	88	1 Marconi Wireless Tel.	71	1 Marconi Wireless Tel.	84
2 Marconi Wireless Tel.	88	2 STC	44	2 STC	61	2 STC	65
3 Sperry Gyroscope	60	3 Sperry	33	3 CSF	37	3 Decca Record	57
4 Metro-Vickers Elec.	22	4 BTH (AEI)	30	4 EMI	32	4 Telefunken GmbH.	43
5 Minister of Supply	19	5 Decca Record	22	5 Sperry	24	5 EMI	36
6 Decca Record	18	6 Bendix Aviation	21	6 Bendix Aviation	23	6 CSF	32
Western Electric	18	Western Electric	21	7 Belling and Lee	22	7 Kelvin and Hughes	31
8 Sadir-Carpentier	17	8 Minister of Supply	19	8 Decca Record	21	8 GEC	29
9 Philips Lamps	16	9 EMI	18	9 Philips Elec. Inds.	19	9 Philips Elec. Inds.	23
10 BTH	13	10 GEC	16	10 GEC	18	10 Bendix	22
11 CSF	11	Sperry Gyroscope	16	BTH	18	Sperry Rand	22
GEC	11	12 Dept. of Commerce, Dir. of the Office of Technical Services of Gilfillen Bros.	15	12 Sperry Rand	15	12 NRDC	18
13 Philco Products	10	Office National d'Etudes et de Recherches Aeronautiques	15	Svenska A.B. Gasaccumulator	15	13 Communications Patents	15
14 Bendix Aviation	8	N.V. Philips	15	Telefunken Ges.	15	14 Belling and Lee	14
15 Hazeltine	7	15 Philco	14	15 Communications Patents	14	Collins Radio	14
Soc. Fr. Radio-Electrique	7	Philips Elec. Inds.	14	16 Metro-Vickers Elec.	12	Cie Fr. Thomson-Houston	14
Submarine Signal	7	17 RCA	12	Minister of Supply	12	17 Hughes Aircraft	13
18 Cossor	6			18 Hazeltine	11	18 Fernseh	12
N.V. Philips	6			19 Antiference	10	19 Atlas Werke	11
20 Patelhold Patentverwertungs u. Elektroholding	5			20 Wolsey Electronics	9	Bush and Rank Cintel	11

Patentee	No.	Patentee	No.	Patentee	No.	Patentee	No.
Pye	5	18 Belling and Lee	11	21 Mullard Radio Valve	11	Minister of Supply	11
22 Antiference	4	Cossor	11	Raytheon Mfg.	11	22 AEI	10
Link Aviation Devices Ind.	4	20 Hazeltine	10	Siemens and Halske	10	BTH	10
RCA	3	21 Metro-Vickers Elec.	8	Philco	8	Meadow-Dale Mfg.	10
25 Aerialite	3	Sperry Gyroscope	8	25 N.V. Hollandse Signaalapparaten	8	25 Avel Corp. Geneva	9
Automatic Signal	3	23 Antiference	7	RCA	7	26 Cossor	8
EMI	3	CSF	7	27 Collins Radio	6	GEC	8
Westinghouse Elec. Inter.	2	25 Belcher (Radio Services)	6	Gen. Precision Lab.	6	Lab. for Electronics	8
29 Curtis Pump	2	Leland Stanford Junior Univ., Board of Trustees of	6	NRDC	6	Pye	8
Hughes and Son	2	Patelhold Patentverwertungs u. Elektroholding	6	Rank Cintel	6	30 Jersey Prod. Res.	7
KLG Sparking Plugs	2	28 Raytheon Mfg.	5			Minister of Aviation	7
Leland Stanford University, Board of Trustees of	2	Seismograph Service	5			Rank Cintel	7
Philco	2	Svenska A.B. Gasaccumulator	5			Socony Mobil Oil Co.	7
Philips Balham Works	2					Svenska A.B. Gasaccumulator	7
Raytheon Mfg.	2						
Soc. Industrielle des Procédés Loth	2						
Sperry	2						

Source: Patent Office, London. This table is based on Group XL(c), Class 40(7) of the Patent Office classification, covering radio navigation, radio-location and aerials. Patent statistics do not provide an accurate indication of R & D activity because firms vary in their patenting policy and patents vary enormously in their value. But over a long period and within a particular sector of industry their usefulness is greater. Firms normally take out far more patents in their home country than abroad. Consequently, the figures for London patents are heavily biased in favour of British firms, and to a lesser extent, of foreign firms with manufacturing subsidiaries in Britain. However, as London is one of the main world centres for patenting, almost all firms selling on the world markets will tend to take out London patents for their most important inventions.

The principal developments since the war have been in three directions. First, the know-how acquired during the war was applied to new civil uses. On the whole, British firms led in this field; but the firms which were prominent in radar during the war (Metro-Vickers and Cossor) were not the leaders in the post-war civil applications. Marconi and Decca, firms only marginally involved in wartime work on radar, took the lead in the development of new types of equipment, such as the Decca marine radar and the Doppler airborne equipment developed by Marconi after feasibility studies by the RRE. Their impressive post-war patenting record, which may indicate very roughly their contribution to new R & D, is shown in Table 4.2, but the reservations in the footnote to the table should be taken into account in interpreting the patent statistics. Nevertheless, their technical leadership was strong enough for them to achieve and maintain a relatively high share of world exports of radar equipment.

Secondly, new military equipment has been developed—early warning systems with long-range capability, low flying interceptor systems and very short wave airborne reconnaissance radar. Thirdly, much more complex control systems have been devised, based on computer links with radar chains. These have civil applications as well as air traffic control. On the whole, the United States has been ahead in these developments and American firms dominate world production and exports.

For the period 1952–61, the percentage of total patents taken out by United States applicants *in London* was 18.4 per cent, but for most electronic categories it was far higher. In radar and navigational aids this proportion was over 30 per cent. American firms with subsidiaries in the UK, such as STC and Sperry, were consistently among the leaders. While the share of French and German firms was relatively small, by the late 1950s Telefunken and CSF were again among the first ten (Table 4.2); both made licensing agreements with American and British firms to try and retrieve their technical lag.

Finally, there have been new developments both in space and defence arising from the wartime work on radar servo-control systems and guidance systems. For reasons of brevity these are not dealt with here, but the most spectacular applications have been in the United States space programme and in communication satellites. With NASA expenditure in the 1960s equivalent to the total R & D expenditure of a large European country, the United States had an unchallenged technical lead in this whole area, which has been maintained into the 1980s despite increased European and Japanese efforts.

Computers[6]

Following the pioneer work on calculating machines by Leibnitz, Pascal, Schickard, Jacquard and others, Babbage began work over a hundred years ago on an 'analytical engine' which already embodied all the main features of the modern computer. Babbage had received one of the first large government development grants, amounting to £17,000 over twenty years for the development of his 'difference engine', but neither this (a special purpose calculating machine) nor the analytical engine (a general purpose machine) was ever completed, because the available components and techniques were inadequate for the purpose.

It is widely believed in Britain and America that the first successful computers

[6] For the early history of computers see Zuse (1961), Watson (1963), Hollingdale and Toothill (1965), Katz and Phillips (1982).

were built in these countries; but in fact the first one, the Z3, was developed by Zuse in Berlin between 1936 and 1941. Zuse began his work while still an engineering student at the Technical High School and much of his early development (the Z1 and Z2) was done with only a few colleagues assisting. Already in 1942 a second Zuse model, the Z4, was used for aircraft design calculations at the Henschel works. It would probably have been in operation still earlier if Zuse had not been called up in 1939 and only released in 1940. The development then had some official support from the German Experimental Aeronautical Institute with some assistance from the Darmstadt and Charlottenburg Technical High Schools.

The Zuse Z3 and Z4 were both electro-mechanical and slow by electronic standards; so also was the first Harvard–IBM computer, the Automatic Sequence Controlled Calculator (ASCC), which was under construction from 1937 to 1944. Multiplication of two numbers took about five seconds on the ASCC and on the Z3, but addition and subtraction took only about 0.07 seconds on the Z3 compared with 0.3 seconds on the ASCC. The total development cost of the German machines was far less than the $400,000 spent on the ASCC at Harvard, or the $500,000 on the later American ENIAC (Electronic Numerical Integrator and Calculator).

Even the first electronic machines were over a thousand times faster than the electro-mechanical ones for both multiplication and addition (Table 4.1). Zuse was also the first to begin work on an electronic computer, in collaboration with Dr Schreyer at Charlottenburg. Special valves were ordered from Telefunken and a bread-board model in 1942 was very promising. But the work was stopped when Schreyer was called up and official support for the project was refused. This and the disruption at the end of the war meant that the lead in computer development passed to the United States and to a lesser extent, Britain.

Wartime applications of embryonic computers had been promoted in both these countries, particularly for code-cracking in Britain (Jones, 1978). In the United States work on the first electronic computer, ENIAC, began at the University of Pennsylvania in 1942 and was completed in 1946. It received financial support from the Army and was mainly designed for calculating trajectories of shells and bombs. It used 18,000 valves, whereas the first Zuse electronic machine would have used only 1,500.

Katz and Phillips (1982), in their fascinating account of the early history of the US computer industry, make particularly interesting comments on the reasons why private funds were not committed to the commercialization of the electronic computer in the early post-war period:

. . . the general view prior to 1950 was that there was no commercial demand for computers. Thomas J. Watson Senior, with experience dating from at least 1928, was perhaps as acquainted with both business needs and the capabilities of advanced computation devices as any business leader. He felt that the one SSEC machine which was on display at IBM's New York offices 'could solve all the scientific problems in the world involving scientific calculations'. He saw no commercial possibilities. This view, moreover, persisted even though some private firms that were potential users of computers—the major life insurance companies, telecommunications providers, aircraft manufacturers and others, were reasonably informed about the emerging technology. A broad business need was not apparent.

From 1945 to 1955, rapid progress was made in solving some of the problems of logic design, memory storage systems and programming techniques. The United States Army, Navy and Air Force, the Atomic Energy Commission and the National Bureau of Standards all placed major contracts for the development of improved computers with universities and with the first firms who began design and manufacture—especially Remington Rand (Univac) and, later, IBM. The major technical advances were made in these large computers and most of the medium and small computers incorporated the developments of their larger brethren scaled down for less complex EDP and scientific applications. Almost all the early demand in the United States was from the military market. Few people then envisaged the large-scale use of computers for data processing, and both government and industry thought mainly in terms of military and scientific applications.

Even after they produced the 650 under the pressures of the Korean War, IBM were still underestimating completely the potential future market. Their Product Planning and Sales Department forecast that there would be no normal commercial sales of the 650, while the Applied Science Group forecast a sale of 200 machines. In the eventual outcome, over 1,800 machines were sold and the 650 became known as the 'Model T' of the computer industry. This indicates very strongly the limitations of theories of market demand leadership for radical innovations and the key role of patient government sponsorship in the early period of radically new technology.

It was perhaps particularly surprising that IBM of all firms did not then appreciate the potential commercial EDP market. The dominant personality in the company was Thomas J. Watson Senior, who was notable for his insistence on product innovation rather than price cuts as the way to enlarge market share, and for his flow of suggestions to the development and engineering departments for improvements in equipment. He seldom admitted any distinction between sales or market research and technical research. Confronted with financial difficulties at IBM when he took over, he borrowed $40,000 and used $25,000 to develop a new tabulator. One of his associates wrote: 'It required a great deal of courage to authorize the tremendous expense for this development and there were many of us who seriously doubted that the customers would stand for the increased rental necessary for the increased complication of the machines; but it is now evident that Mr Watson had correctly estimated the final result.' It took four years to develop the machine. He continued to insist on the importance of new product development, particularly during the depression of the 1930s. 'We had some new machines and ideas to give our salesmen. . . . If we had to depend on the line we had five years ago, it would have been a different story.' It was also very characteristic of him to insist on the importance of education. Few companies can have given so much attention to the selection and training of their own employees and to the training of their customers. The IBM Department of Education in 1956 (that is, before their big expansion in electronic computers) had a budget of $14 million—nearly 3 per cent of turnover.

Powers/Remington Rand was the technical leader and had a stronger R & D effort than IBM (Hoffman, 1976), but by the 1930s IBM had by far the largest market share based on the success of its world-wide service and sales organization and field-force of engineers.[7] It also had manufacturing subsidiaries in France, Germany and several other countries. It had over three-quarters of the punched-card

[7] IBM's gross income grew from $116 million in 1946 to $696 million in 1955, and employment from 22,000 to 41,000. By 1964 gross income reached $3,239 million and employment was 149,000. In 1971 gross income was $8.273 million and there were 265,000 employees.

market, whose users were the obvious outlet for EDP. It had fairly important R & D facilities before the Second World War, and was taking out more patents in London in the field of calculating machinery than any other firm (Table 4.3). (Until 1949 the British Tabulating Machinery Company, BTM, had an exclusive franchise for all IBM products in Britain, and took out patents on IBM inventions. It did not conduct its own R & D.) Not only IBM, but also other American companies, such as NCR, have had a very strong position in patents in this area for over thirty years, even as measured by patents taken out in London (Table 4.3). US companies consistently accounted for about half of all *London* patents for calculating machinery for about fifty years.

The existing data processing equipment in the early 1950s (mainly punch-card calculators and tabulators) was profitable both for IBM and for British companies such as BTM and Power-Samas. Consequently, there was some reluctance to make this equipment obsolete by introducing electronic computers. Many firms in this kind of situation have failed to innovate, or left it too late. What was remarkable about IBM was the speed with which it recovered from a situation in which it was in danger of falling behind and embarked on the large-scale development and manufacture of the new products once their advantage had been demonstrated. This change was also associated with a change in management and with the settlement of an anti-trust suit brought by the Department of Justice against IBM over its dominant position in the punched-card market.

The story has been told by Watson's son (Belden and Belden, 1962):

One of the most exciting chapters in IBM's post-war history has to do with large-scale electronic computers and data processing. Many very large engineering computational jobs and a fair number of accounting applications were being hampered by the slowness of the calculating machines available in the later 1940s. However, Drs Eckert and Mauchly of the Moore School at the University of Pennsylvania had built a large electric computer—the Eniac—for the Army to make ballistic curve calculations. Many of us in our industry, including me, had seen the machine, but none of us could foresee its capabilities. Even after Eckert and Mauchly left the Moore School and began privately to manufacture a civilian counterpart to the Eniac—the Univac—few saw the potential.

The Company was finally absorbed by Remington Rand in 1950, and soon had installed several machines in the US Government, including one in the Census Bureau which replaced a number of IBM machines.

During these really earth-shaking developments in the accounting machine industry, IBM slept soundly. We had put the first electronically-operated punched card calculator on the market in 1947. We clearly knew that electronic computing even in those days was so fast that the machine waited 9/10 of every card cycle for the mechanical portions of the machine to feed the next card. In spite of this, we didn't jump to the obvious conclusion that if we could feed data more rapidly, we could increase speeds by 900 per cent. Remington Rand and Univac drew this conclusion and were off to the races.

Finally we awoke and began to act. We took one of our most competent operating executives with a reputation for fearlessness and competence and put him in charge of all phases of the development of an IBM large-scale electronic computer. He and we were successful.

How did we come from behind? First, we had enough cash to carry loads of engineering, research and production, which were heavy. Second, we had

Table 4.3   Patents taken out in London for computing, calculating and registering machinery

1931–8	No.	1939–46	No.	1947–50	No.
1 BTM[a]	137	1 BTM[a]	88	1 NCR (US)	59
2 NCR (US)	115	2 STC	41	2 BTM[a]	37
3 BTH	57	3 Bendix		3 Bendix	
4 Burroughs	51	Aviation	31	Aviation	20
5 Siemens	35	4 Remington		4 BTH	17
6 Ericsson		Rand	28	STC	17
Telephones	31	5 NCR (UK)	25	6 Remington	
7 Siemens and		BTH	25	Rand	16
Halske	30	7 NCR (US)	24	7 Jack & Heintz	12
8 Landis & Gyr	25	Kappelia	24	Landis & Gyr	12
9 Automatic		9 S. Smith	22	Sperry	
Electric Co.	23	10 English		Gyroscope	12
10 Mercedes Buro		Electric	17	10 Lapco Soc.	
maschinen-		11 British Adding		Anon.	11
werk	22	and Calculat-		11 Allen-Wales	
11 Electroflo		ing Machines	15	Adding Machine	
Meters Co.	18	Kooperativa		Corp.	10
AEG-		Forbundet		12 W. & T. Avery	9
Fahrkarten-		Forening	15	GEC	9
drucker		Heenen and		IBM	9
Ges.	18	Froude	15	Sheffield	9
13 F. Krupp	17	14 IBM	14	Sigma Instr.	9
14 Etab. E.		15 Mercedes Buro-		Westinghouse	9
Jaeger	14	maschinen-		18 Brown & Sons	8
GEC	14	werke	13	Minister of	
16 NCR (UK)	13	Molins		Supply	8
17 Remington		Machine	13	Union	
Rand	12	W. & T. Avery	13	Totalisator	8
Bendix		18 Bell Punch	12	21 Kapella	7
Aviation	12	19 Marconi's		Marconi's W. T.	7
19 Accounting and		W. R.	11	Power Jets	7
Tabulating	11	20 A. B. Svenska		SACMA	7
Dunlop Rubber	11	Kullager-		25 Bell Punch	6
21 Askania Werke	10	fabriken	10	Bryant Chucking	
STC	10	Landis & Gyr	10	Grinder	6
23 Hasler	9	22 Metr. Vickers	9	Clayton	
G. Kent	9	Barnet		Dewandre	6
Kooperative		Instruments	9	Electroflo	
Forbundet		Sheffield	9	Meters	6
UPA	9	Ferranti Ltd	9	Eng. Electric	6
IBM	9	26 Burroughs	8	Felt & Terrant	6
Lamson		Brown & Sons	8	S. Smith	6
Paragon	9	28 Underwood		TIM	6
28 Inter. GEC	8	Elliott Fisher	7		
C. P. Goerz		Dunlop Rubber	7		
Optische		Vickers			
Anstalt	8	Armstrong	7		
J. Bradbury	8	Electroflo			
		Meters	7		
		Addressograph			
		Multigraph	7		
		Hughes & Son	7		

[a] In this period BTM took out patents on IBM inventions.

1951–4	No.	1955–8	No.	1959–62	No.
1 NCR (US)	76	1 IBM	177	1 IBM	246
NRDC	76	2 NCR (US)	111	2 STC	93
3 BTM	72	3 NRDC	103	3 ICT	77
4 Power-Samas	36	4 STC	97	4 NCR (US)	69
5 STC	32	5 BTM	45	5 NRDC	51
6 IBM	29	6 Landis & Gyr	31	6 Sperry Rand	48
7 BTH	28	7 Burroughs	30	7 Philips	43
8 Landis & Gyr	24	8 Bendix		8 AEI	37
9 Bell Punch	22	Aviation	26	9 Western Elec.	36
10 Bendix		9 Power-Samas	24	10 Burroughs	34
Aviation	18	10 Western		11 EMI	32
Cie des Machines		Electric	23	12 GEC Ltd	31
Bull	18	11 Kienzle		13 Kienzle	
12 Soc.		Apparate	22	Apparate	22
d'Electronique		12 BTH	18	14 GE	20
et d'Auto-		Sheffield	18	15 Olympia	
matisme	17	Signal Instr.	18	Werke	18
Sheffield	17	15 Foxboro	17	16 Landis & Gyr	17
14 Remington		16 Brunsviga		17 Rolls Royce	16
Rand	15	Maschinen-		18 Sheffield	15
15 Metro-Vickers	13	werke	16	19 British	
16 A. B. Atvida-		17 GE	15	Telecommun.	
bergs Facit	10	18 Philips	13	Res.	14
N. V. Philips	10	Machines Bull	13	20 Ericsson	
S. Smith	10	20 W. & T. Avery	12	Telephones	13
Western Electric	10	EMI	12	Hensoldt &	
20 Burroughs	9	GEC	12	Sohne	13
Licentia Patent-		Mullard	12	22 Bell Punch	12
Verwaltungs		RCA	12	Machines Bull	12
Ges.	9	25 Felt & Tarrant	10	Ferranti	12
Westinghouse	9	Sangamo		Société	
Gerrard		Weston	10	d'Electronique	
Ticket M.	9	Société		et d'Auto-	
23 Dunlop		d'Electronique		matisme	12
Rubber	8	et d'Auto-		26 English Elec.	11
GEC	8	matisme	10	Magnavox	11
Philips Elec.	8	28 Bell Punch Co.	9	Siemens & Halske	11
Union		Kelvin & Hughes	9	Westinghouse	
Totalisator	8	Olivetti	9	Brake & Signal	11
27 W. & T. Avery	7	Siemens & Halske	9	Veeder-Root	11
Eckert-Mauchly	7	S. Smith & Sons	9		
Ferranti	7				
Foxboro	7				
Rolls Royce	7				
Sigma Instr.	7				
Standard Gage	7				
Totalisators	7				
Verder-Root	7				

Source: Patent Office, London.

a sales force which enabled us to tailor our machine very closely to the market. Finally, and most important—we had good company morale. All concerned realized that this was a mutual challenge to us as an industry leader. We had to respond with all that we had to win, and we did.

By 1956, it became clear that to respond rapidly to challenge, we needed a new organization concept. Prior to the mid-1950s the company was run essentially by one man—T. J. Watson, Sr. He had a terrific team around him, but he made the decisions. If we had organization charts, there would have been a fascinating number of lines—perhaps thirty—running into T. J. Watson.

In the early 1950s the demands of an increasing economic pace and the Korean War were calling for more rapid action by IBM at all levels than our monolithic structure was able to respond to adequately. Increasing customer pressures—plus a few more missed boats of lesser consequence than the Univac situation—forced us to decide on a new and vastly decentralized organization.

Here, we hope we responded a little more rapidly than we did in the case of Univac. The new organization was 180 degrees opposed to the old in fundamental concept, but we made the move.

In late 1956, after months of planning, we called the top one hundred or so people in the business to a three-day meeting in Williamsburg, Virginia. We came away from that meeting decentralized.

The major innovations in electronic digital computers after ENIAC and EDVAC (University of Pennsylvania) were made by Univac (Remington Rand) and MIT rather than IBM (magnetic core memory stores, machine translation of instructions, magnetic drums and discs). Hoffman (1976) has shown in his study of R & D strategies in the computer industry that IBM tended to follow a strategy of rapid imitation, picking up the most important scientific and technical advances from universities and from competitors. In the early days IBM hired von Neumann, one of the leading mathematicians from the EDVAC team, and also received important help from Sperry Rand for their 650 series, first sold in 1954. In the 1950s, government-contract R & D accounted for about 60 per cent of total expenditure at IBM; now private venture expenditure accounts for the greater part of the (much larger) total. By the 1960s, about 15,000 people were engaged on research and development, about 90 per cent of them in eighteen development laboratories (thirteen in America and five in Europe). Some very large computers embodying more advanced features were built on contract for government agencies, but the highly successful transistorized 1401 series (1960), the 360 series (1965) and the 370 series (1971) were private venture projects.

The launching of the 360 series is a particularly interesting illustration of the change in scale of innovative work in this industry compared with the early wartime and post-war computers. The total development costs including software have been estimated at $500 million. It was preceded by the failure of IBM in the larger computer systems with the STRETCH machine, on which it lost $20 million. STRETCH did not meet the promised specification and the price had to be reduced to $8 million from $13.5 million, at which level it sold at a loss and very few were sold. The loss would have been much greater but for government support with the development costs.

There was a strong temptation in the early 1960s to stay with the highly successful 1401 series (General Products Division of IBM) and the 7000 series (Data Systems Division). However, these were overlapping increasingly in range and each Division had its own development programme for extending the range. After sharp internal power struggles and after several new development projects had been killed off, including one based at IBM's Hursley Laboratories in England, Learson and Evans, now in charge of development at Data Systems, decided to concentrate on a completely new compatible series and not to proliferate the existing lines. The new series should cater for both business and scientific users. They also decided to use the new hybrid integrated circuits which were just becoming available. 'It was roughly as though General Motors had decided to scrap its existing makes and models and offer in their place one new line of cars, covering the entire spectrum of demand, with a radically redesigned engine and an exotic fuel.' (Wise, 1966.)

This not only involved over-ruling competing development projects, and concentrating on the NPL (later 360) series. It also involved a huge programme of plant expansion (five new plants) and IBM's entry into circuit manufacture. Between 1964 and 1967 IBM budgeted $4.5 billion in addition to R & D costs for investment in rental machines, plant and equipment.

The risks involved in launching the series were very great and were recognized, but the critical factor in the decision seems to have been the view that if they did *not* launch a new series they risked a serious decline in growth rate and erosion of their position as market leaders, particularly in larger computers. Evans' justification of this course of action was echoed almost word for word in relation to a similar crucial decision by Rolls-Royce to launch a new generation of aero-engines (the Spey). He said that the 360 series was a lot less risk than it would have been to do anything else or to do nothing at all (Wise, 1966).

Altogether, over a dozen other American firms entered the EDP market in addition to IBM, but none of them made any profits from EDP computers before the 1960s and, according to most estimates, hardly any did so later. As with colour television, RCA had to lay out over $100 million and both they and General Electric withdrew from the market in the early 1970s after a heavy investment. In the case of General Electric this included an attempt to establish a strong foothold in Europe by the purchase of the main French manufacturer, Bull. While IBM succeeded in maintaining its overall market dominance throughout the 1970s, it was successfully challenged in special sectors of the market, such as mini-computers and the largest systems. The advent of the micro-computer presented a more serious threat, but once again IBM was able to react through the introduction of its own new competitive products with extraordinary speed. Nevertheless, a number of new firms and products have been able to enter the market, such as Sinclair and Torch in the United Kingdom and many others in the United States and Japan (Sciberras *et al.*, 1978).

Computers in Britain

In Britain, electronic computer development work began in 1946–7, after British scientists and engineers had visited the first American installations. The earliest work was at London, Manchester and Cambridge Universities, and at the Elliot Brothers' (later Elliott Automation) Research Laboratory on digital computing for fire control for the Admiralty. Shortly afterwards the National Physical Laboratory began cooperation with English Electric on the ACE electronic computer, which was completed in 1950. Manchester University, assisted by Ferranti, completed its first electronic models in 1948–9, and the Cambridge EDSAC was also working in 1949. The more ambitious Manchester Automatic Digital Machine (MADM) was completed in 1950 and ten machines were ordered by the National Research and Development

Corporation (NRDC) from Ferranti for the Mark I, based on MADM. In the next ten years the NRDC gave financial backing to several more computer projects, including the Elliott 401, the Ferranti Pegasus, and later the EMIDEC 2400 and the ATLAS. The NRDC also held a number of key university and government patents for the Manchester (Williams) tube storage system (superseded by the core system). Because of its patent holding the NRDC organized a patent 'pool' for those firms engaged on digital computers, which still exists.

However, although the NRDC provided an important stimulus and had a computer committee which already foresaw some of the business EDP possibilities as early as 1949, its total outlay on development for the first ten years was less than a million pounds,[8] and it did not attempt to rationalize the industry or otherwise use its influence to bring together the many firms which became involved.

Other important work was done on a private venture basis. The first straight commercial EDP work in Britain was done by a LEO computer in 1951 on bakery costing applications (and was done weekly for fifteen years thereafter). This was also the first computer to do payroll calculations, although the magnetic tape available at the time was not really suitable. Work was begun at Lyons as early as 1947 under Mr Thompson who, after a preliminary study with Cambridge University Mathematical Laboratory, used their consultancy services and experience with EDSAC to launch LEO (Lyons Electronic Office). The total cost of development from 1947 to 1953, including the consultancy, was £129,000. However, a delivery to an outside customer was not made until April 1958.

After its forty-year connection with IBM was severed in 1949, BTM (British Tabulating Machines) also designed and developed a small machine, the Hollerith 'HEC', using a magnetic drum store based on early joint work at Birkbeck College. The first deliveries were in 1955, and in 1956 BTM formed a joint company with GEC to develop and manufacture that next generation of computers. This company developed the transistorized (ICT) 1301, of which deliveries began in 1962. Meanwhile BTM had merged with Power-Samas in 1959 to form ICT, which subsequently absorbed the EMI computer division (1962) and the Ferranti EDP computer division (1963). By these and other mergers, such as the formation of English Electric-Leo-Marconi in 1964, the number of firms developing and manufacturing EDP computers in Britain was reduced from ten to three (excluding Honeywell and IBM), and the new Ministry of Technology used its influence a few years later to reduce this to one, International Computers Limited. Most of the original manufacturers had succeeded in developing a successful computer, but few models had sales going into three figures, or even over fifty. This created very serious problems for the recovery of heavy development and technical service costs, which are discussed in the concluding section of this chapter and again in chapter 6.

Some contrasts between British and American development stand out. First, many American government departments, particularly service departments, placed development contracts and hardware orders with universities and industry between 1946 and 1955. In Britain the only order for a large machine came from the Royal Aircraft Establishment and the National Physical Laboratory for the English Electric DEUCE. The first orders from government departments for data processing machines did not come until much later on. This was despite the efforts of the firms to interest the Services in ordnance and stock control applications, which were rapidly taken up in the United States.

---

[8] There was additional expenditure on orders to firms.

Secondly, although early prototypes were developed by several firms and universities fairly quickly, and although a number of firms made successful 'first generation' valve machines in the 1950s, the British industry lagged behind the American in developing the 'second generation' of transistorized machines for commercial EDP. This was despite the success of computers designed primarily for scientific purposes, or process control, such as the Elliott 803 (1959). The distinction between the 'scientific' and the commercial 'EDP' markets is important. Sometimes they may use the same machines, although with different configurations and peripherals. But, whereas the 'scientific' customer usually knows a good deal about the machine and can do a lot of his own 'software' and maintenance, the 'commercial' customer usually needs a great deal of training, advice and assistance from the manufacturer (Brock, 1975).

The field force in the EDP market must be much larger and this is an important scale factor advantage for large firms such as IBM. Firms which are successful in one market will not necessarily be successful in the other. Some of the British manufacturers tended to regard EDP customers as though they were scientific customers and British development work lagged behind in 'software' and in the peripherals for commercial users. The heavy costs of successive new developments, and still more the cost of selling and maintaining equipment and training staff, in relation to the relatively small sales achieved, made it an unprofitable venture for the firms involved.

This was even more true in Continental Europe. Zuse did succeed in starting work on a small scale and delivering his first post-war computer to a commercial customer in 1953 (the Z5), and also developed a valve model, the Z22 (1956–8), and a transistorized model, the Z23 (in 1958–61). Other European firms did not begin delivery of computers until 1958 (Siemens) and 1959 (Bull) and did not succeed in maintaining independent development and manufacture, although Siemens is renewing its effort with strong government backing. Zuse also, although successful in developing and selling scientific computers on a small scale, was taken over by Brown Boveri in 1964. Only in very small computers did European firms succeed in retaining a competitive position in the world market. In medium-sized and large computer systems government policies of preferential purchases and R & D subsidies were necessary to sustain the independent existence of non-American firms in the UK, West Germany and Japan.

Electronic components[9]

The development process in the electronic capital goods industry consists largely of devising methods of assembling components in new ways, incorporating new components to make a new design, or developing new components to meet new design requirements. This is not quite so simple as it may sound. There are more than a hundred thousand different components in a large computer, more than a million in a big electronic telephone exchange and ten million in the Apollo rocket system. There must be close collaboration in design work between end-product makers and component makers, and in the more complex products there must be sophisticated programming of component supply and sub-assemblies, and of testing arrangements.

If component makers succeed in developing new products with outstanding

[9] For a thorough treatment of component innovation see Golding (1972), OECD (1968), Tilton (1971), Sciberras (1977), Braun and MacDonald (1978) and Dosi (1982).

improvements in performance, this will most quickly benefit those who are in close touch with them. Before the Second World War European firms were probably not at any disadvantage here: some of the most important advances in valves and tubes were made in Europe by Philips, Telefunken,[10] GEC, AEI (BTH) and Marconi, as well as by European universities. Since the war the position has changed completely: almost all the important inventions and innovations in components have been made in the United States, and there has been a lag of one to four years before manufacture began in Europe, often by American subsidiaries (Table 4.4). The major breakthrough was the invention of the transistor at Bell Laboratories and associated semi-conductor devices; but there have also been revolutionary improvements in the design and performance of most other components, including the so-called 'passive' components (such as resistors and capacitors).

These innovations and improvements depended on advances in fundamental research as well as in development. The discovery of the transistor was made in the laboratories of a firm which has always spent heavily on *both* research and development. Almost all firms in the industry do a lot of development, but few do much research. This is partly a question of size. Development is directed to the introduction of a definite product; research is a more open-ended process, with bigger risks and a longer time perspective. It is firms with large resources and a wide product range who tend to devote a higher proportion of their total R & D expenditure to research. Even these seldom devote more than about 25 per cent of their total outlay to both fundamental research and applied research, allocating the remainder to development. Public corporations tend to allocate a higher proportion to research.

Component-makers tend to do more research than equipment-makers. Some of the large American electrical companies consider it very important that they should do some fundamental research in their main fields of interest and have gone to some trouble to simulate an academic environment in their laboratories. Among these Bell Laboratories has been ourstanding. The company estimated total R & D expenditure on transistors and transistorized equipment at Bell Laboratories at £2.7 million up to 1953, at £28 million up to the end of 1960, and £57 million up to September 1964. At this time there were about half a dozen American electrical companies, any one of which was spending nearly as much on electronics R & D as the German Federal Republic or France. They could afford to behave like the government of a medium-sized country or a public corporation in their endowment of research. Total R & D expenditure of the leading companies is shown in Table 4.5. The figures are for company-financed R & D only. Many of the leading aircraft and electronics companies would be receiving Federal R & D contracts at least as large as their company-financed expenditure. These estimates cover all R & D, much of which is non-electronic, the proportion varying within each company. In the 1960s and 1970s, probably only Philips or Siemens among European firms could afford to match this scale of expenditure. But the rationalized GEC–AEI–English Electric has been in this league since 1968, along with several Japanese companies. The twenty leading Japanese firms (in terms of R & D expenditures in 1979) included half-a-dozen electronic firms which were in the American league (Table 4.5), although lacking the heavy US contribution from government for military electronics.

The Bell basic patents in transistors were made available to all-comers on payment

[10] For example, the pentode valve was developed at Philips and the hexode by Steimel at Telefunken.

Table 4.4　First commercial production of various electronic components

Component	United States		United Kingdom		Italy		Western Germany	
	Date	*Firm*	*Date*	*Firm*	*Date*	*Firm*	*Date*	*Firm*
Silicon mixer diode (point contact)	1938	GE	1939	BTH[a]			1943	Telefunken[a]
Germanium diode	1941	GE[a]	1942	BTH			1943	Telefunken[a]
Point-contact germanium transistors	1951–2	Bell[a], WE	1953	AEI, STC[b]			1954	Telefunken[a]
Silicon junction diode	1952	Bell[a], WE	1954	Ferranti	1959	SGS	1953	Valvo
Germanium junction transistor	1951	GE, RCA	1953	Mullard[b]	1954	Philips		Telefunken
Silicon-grown junction transistor	1954	TI[a]	1957	TI[b]	1958	Philips	1959	Valvo
Alloy-diffused germanium transistor	1955	Bell, WE	1958	Mullard[b]	1960	Philips	1960	AEG[a]
Silicon controlled rectifier	1956	GE[a]	1957	AEI[a]	1960	SGS	1957	Siemens[a]
Silicon power rectifier	1955	GE[a]	1957	AEI,[a] STC			1958	AEG[a]
Germanium power rectifier			1954	AEI				Siemens
Diffused (mesa) silicon transistor	1958	TI[a]	1959	TI, Ferranti	1959	SGS		Siemens
Diffused (mesa) germanium transistor	1956	TI	1958	TI				Valvo
Planar transistors	1960	Fairchild[a]	1961	Ferranti	1962	SGS[b]	1963	Intermetall
Epitaxial transistor	1960	WE	1962	TI, STC			1963	Valvo
Integrated monolithic microcircuits	1961	TI[a]	1963	TI[b] Ferranti[a]	1966	SGS[b]	1964	Telefunken

[a] Own R & D
[b] Under licence

SGS = Societa Generale Semiconduttore (partly owed by Olivetti, Telettra and Fairchild); GE = General Electric; WE = Western Electric (WE forms the manufacturing function and Bell Labs the research function of AT&T = American Telephone and Telegraph); RCA = Radio Corporation of America; TI = Texas Instruments; BTH = British Thomson Houston (later part of AEI and of GEC/AEI); STC = Standard Telephone Company (British subsidiary of ITT); Valvo and Mullard are subsidiaries of Philips.

Source: US Dept. of Commerce, September 1963, p. 150; and information from firms.

Table 4.5a Leading US firms in company-financed R & D expenditure ($ million) 1980 (excluding Federal contracts)

Rank	R & D expenditures	R & D as % of sales
		%
1 General Motors	2224	3.9
2 Ford	1675	4.5
3 IBM	1520	5.8
4 American Telephone and Telegraph (AT&T) (including Bell and Western Electric)	1338	0.8
5 Boeing	767	8.1
6 General Electric	760	3.0
7 United Technologies	660	5.4
8 Eastman Kodak	520	5.3
9 International Telephone and Telegraph (ITT)	504	2.7
10 Exxon	489	0.5
11 Du Pont	484	3.5
12 Xerox	434	5.3
13 Sperry Rand	337	6.2
14 Dow Chemical	314	3.0
15 Honeywell	295	6.0
16 Hewlett-Packard	272	8.8
17 Minneapolis Mining and Manufacturing	283	4.6
18 Chrysler	278	3.0
19 Merck	234	8.6
20 Johnson and Johnson	233	4.8

Note: The exchange rate for the Yen in 1980 was 227 Yen per US Dollar.
Source: *Business Week*, 6 July 1981.

Table 4.5b Twenty leading Japanese firms in expenditures on R & D in 1979[a]

Rank	Billion Yen	Ratio to Total Sales
		%
1 Toyota Motor	104.0	3.7
2 Hitachi	98.7	5.8
3 Nissan Motor	90.0	3.3
4 Toshiba	69.0	4.8
5 Matsushita Electrical Ind.	50.0	2.9
6 Nippon Electric	43.0	6.0
6 Mitsubishi Electric	43.0	4.0
8 Mitsubishi Heavy Ind.	38.2	2.8
9 Honda Motor	38.0	3.6
10 Sony	32.8	7.0
11 Fujitsu	30.5	6.1
12 Nippon Steel	27.0	1.0
13 Toyo Kogyo	20.5	2.5
14 Nippondenso	20.5	4.5
15 Takeda Pharmaceutical	20.1	4.8
16 Fuji Photo Film	18.8	6.0
17 Isuzu	18.6	2.9
18 Bridgestone	18.0	4.1
19 Kobe Steel	17.7	1.7
20 Tokyo Electric Power	15.2	0.7

[a] Financial year.
Source: Survey conducted by the Nihon Keizai Shimbun covered 1,170 firms of which 1,015 in manufacturing and 155 in non-manufacturing.

Table 4.6a Major product innovations in the semiconductor industry since the integrated circuit

Innovations	Firm	First commercial production
MOS transistor	Fairchild	1962
DTL integrated circuit	Signetics	1962
Gunn diodes	IBM	1963
Light-emitting diodes	Texas Instruments	1964
TTL integrated circuit	TRW	1964
MOS integrated circuit	General Microelectronics	1965
	General Instruments	
Magnetic bubble memory	Western Electric	
MOSFET (MOS field-effect transistor)	Western Electric	1968
	Philips	
Schottky TTL	Texas Instruments	1969
CCD (charge coupled device)	Fairchild	1969
Complementary MOS	RCA	1969
Static RAM	Intel	1969
Silicon-on-sapphire (SOS)	RCA	1970
P-MOS		1971
3-transistor cell dynamic RAM (1K bits)	Intel	1971
CMOS		1971
Microprocessor	Intel	1972
I^2L integrated circuit	Philips	1973
1-Translator cell dynamic RAM (4K bits)	Intel	1974
VMOS integrated circuit	AMI	1975
C^2L integrated circuit		1976
MNOS		1976
Micro-computer (8048)	Intel	1977
V-MOS	Mitsubishi	1978
64-K bits memory	Fujitsu	1978

Source: Dosi (1981).

Table 4.6b Major process innovations in semiconductor industry

Innovation	Firm	Date of development
Single crystal growing	Western Electric	1950
Zone refining	Western Electric	1950
Alloy process	General Electric	1952
3-5 Compounds	Siemens	1952
Jet etching	Philco	1953
Oxide masking and diffusion	Western Electric	1955
Planar process	Fairchild	1960
Epitaxial process	Western Electric	1960
Plastic encapsulation	General Electric	1963
Beam lead	Western Electric	1964
Dielectric isolation	Motorola	1965
Collector diffusion isolation	Western Electric	1969
Ion implantation	Mostek	1970
Self-aligned silicon gate	Intel	1972
Integrated injection logic	Philips	1973
Vertically oriented transistor	AMI	1975
Double polysilicon process	Mostek	1976
E-beam mask projection		1976
Plasma nitride processing		1976
Automatic bonding on 'exotic' (35 mm film) substrate	Sharp (Japan)	1977
Vertical injection logic	Mitsubishi	1978

Source: Dosi (1981).

of a $25,000 advance royalty, partly because of an anti-trust suit (filed by the Department of Justice in 1949 and finally settled by a consent decree in 1956). But licensing arrangements subsequent to the decree were negotiated individually with each company. Those who could offer know-how in return obtained more favourable terms, often paying only the $25,000 lump sum.

Between 1952 and 1963 the Western Electric Co. (Bell) received over £9 million in total income for royalties from companies all over the world, excluding cross-licence benefits, and of this the Company estimated that over £3 million was attributable to transistors. By far the greater part of this income came from companies outside the United States (£556,000 from licensees having their principal office in the United Kingdom). United States concerns were able to use Bell patents prior to the consent decree of 1956 royalty-free, but pay royalties (varying in scale with the individual agreement) on patents issued since the decree.

Total expenditure on the negotiation and administration of licence agreements was £6.3 million from 1952 to September 1964, of which £1 million was for transistors and transistorized equipment. Cross-licence benefits in respect of agreements involving the basic transistor patent were estimated at £2.6 million. Only £4,000 of this was attributable to UK companies. Thus licensing income did not lead to the direct recovery of more than a small part of Bell's R & D costs.

Technological change continued at a rapid rate throughout the 1960s and 1970s and, as a result of their technical lead, other American component firms, in addition to Bell, enjoy a considerable royalty and know-how income from the licensing agreements which they have concluded with European firms. The planar technology patents of Fairchild were particularly important, and in addition to European firms, leading American firms, such as Texas Instruments and ITT, made licence agreements with Fairchild to obtain this technology. There were, it is true, a few component developments in Europe which were licensed to the US: for example, the Lucas development work in industrial semiconductors, which began in 1954, resulted in the successful development of high voltage devices for ignition systems, which were licensed to Delco in the US. Production of the Gunn diode was first launched in the UK and Siemens licensed Westinghouse for their ultra-pure silicon process. But, in total, American companies have a substantial positive balance in 'technical payments' in the component field; this reflects their lead in most areas of new component development since the war. Developments in the microprocessor field in the 1970s and 1980s generally enhanced this lead with US firms accounting for almost all the major innovations (Table 4.6). Japanese firms were extremely quick with imitation, application and improvement inventions and innovations, but had not overtaken the US firms in radical innovations.

Diffusion of technical know-how does not simply depend on ability to pay. It owes a great deal to personal contacts and discussion, or to the movement of people—and here American firms enjoyed a major advantage. The Research Director of Texas Instruments came from Bell and so did other key personnel in the American semiconductor industry. As Tables 4.4 and 4.5 suggest, Texas Instruments, Fairchild and Intel have made an especially important contribution to the advance of component technology since the original Bell innovations. Golding has documented in detail the great importance of the movement of key R & D personnel in the development of other firms (Figure 4.1). American firms in 'Silicon Valley' and elsewhere not only benefited from this but also from the close contacts with the equipment-makers and universities, such as Stanford. These advantages of recruitment and close contact and communication have led many

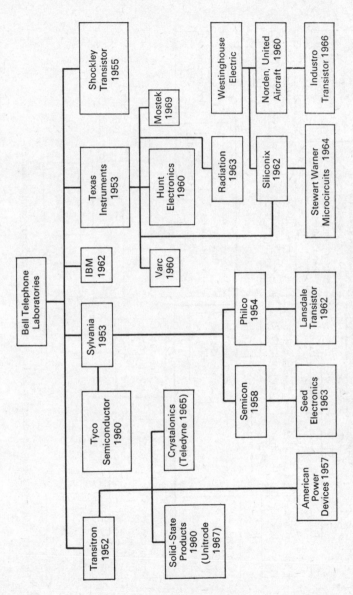

Fig. 4.1a   Spin-offs from Bell Telephone Laboratories and Westinghouse Electric

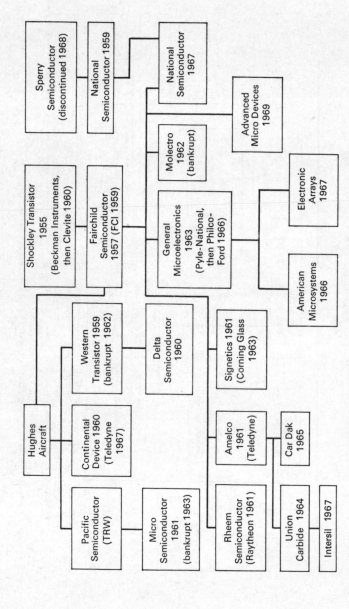

Fig. 4.1b   Spin-offs from Shockley Transistor, Hughes Aircraft and Sperry Semiconductor

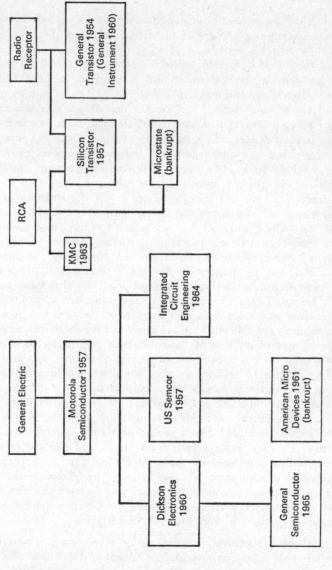

Fig. 4.1c Spin-offs from General Electric, RCA and Radio Receptor. *Source*: Golding (1972)

of the leading European and Japanese firms to set up subsidiaries in the American electronics industry, often through the purchase of small or medium-sized US firms.

After the major discoveries at Bell, the US government took a hand in advancing component technology. Since 1950 it has financed R & D on a large scale, it has helped the principal firms to enlarge their capacity, and it has placed large orders for devices. A study by Little (1963) found that:

Due to its considerable interest in semiconductors and particularly in transistors, the government has throughout the 1950s tried to stimulate the development of improved types. Around the middle of 1950 they were convinced transistors were needed for future military equipment so they accelerated the production investment to provide developmental and production facilities for making certain types which were considered to be desirable for future military electronic equipment. Thus, whereas during most years in that decade government funding for research, development and industrial preparedness work ran at the rate of four to eight million dollars each year, in 1956 a major additional appropriation became available and was channelled into industrial preparedness studies on transistors. Resulting in a $14 million additional expenditure, the contracts granted extended over the next two or three following years.

The contracts for a total of thirty different types of germanium and silicon transistors were placed with about one dozen of the major semiconductor companies, and this helped some of these to gain a foothold in the industry. In many cases, this investment was matched by similar amounts of capital equipment or plant space supplied by the contracting companies. Requirements of the contract included the delivery of about three thousand transistors of each type to be made on production lines capable of producing that many per month. Thus a total potential capacity of over a million transistors a year was created. Since at that time transistors were manufactured at yields as low as 5 per cent to 15 per cent, it is not unrealistic to assume that a productive capacity eventually capable of tens of millions of transistors was created by these contracts, and this at a time when the total unit shipments of finished transistors by the industry ranged from 14 million units in 1956 to 28 million in 1957. Here although the individual companies certainly paid for the plant space, almost all engineering design and development was paid for by the government.

An additional impetus to the industry resulted also from the clear delineation of the specifications required which predicted certain types that were likely to be in major military usage a few years hence, and this stimulated many companies including a number of those which had not participated in this earlier windfall to develop types at least as good, preferably better, than the ones under the government contract in order to gain back this potential business. Possibly the present overcapacity within the industry was created during this period, which resulted in major price declines beginning in 1960.

These conclusions were confirmed by the results of Golding's research (1972) on economies of scale in the semiconductor industry. He found that dynamic scale economies associated with the learning curve were particularly important for new devices because of importance of the 'yield' phenomenon in semi-conductor manufacture. The difficulties of maintaining very high purity and cleanliness in the manufacture of microscopically small components mean that yields initially may be as low as 5 to 10 per cent. Obviously, very great economies can be secured by improving this yield, and to a considerable extent this proved to be a function of production volume. Similar phenomena have been found in other industrial processes subject to rapid technical change, and were first documented in connection with the aircraft industry. Golding (1972) estimated that a unit cost reduction of around 20 to 30 per cent was associated with cumulative doubled quantities (Andress, 1954; Sturmey, 1958; Beloff, 1966). These are twice as great as those estimated for aircraft production in the Plowden Report (1965) (see Figure 4.2).

These 'dynamic' economies of scale arise from adaptive learning of the labour force and management engaged in the production process. They are additional to, but intimately related to improvements arising from R & D, and from the normal 'static' economies of scale—reduction in unit costs arising from the spread of fixed costs over a larger production and sales volume.

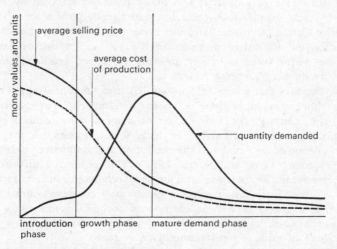

Fig. 4.2   The pattern of a typical semiconductor product cycle. *Source*: Golding (1972).

Evidently, the increases in production volume for new devices arising from US government orders gave an enormous competitive advantage to US manufacturers, compared with European manufacturers, who were faced with much smaller demand and much greater uncertainty. This has been reflected in the dominance of the world market by the US manufacturers, the high proportion of semiconductor imports into Europe from the US, the establishment of US subsidiaries in several European countries, and the great difficulties experienced by all but the largest European firms in survival in this industry. There is some difference of opinion on the pricing strategy followed by the leading innovative firms in the US semiconductor industry. Sciberras (1977) has maintained that they were prepared to bring prices below costs of production for some time in the early stages, confident that if they achieved a large market, dynamic economies of scale would enable them to recoup their losses. He believed that this was quite consistent with a profit maximization hypothesis since in the long run this strategy would erect effective barriers to entry. Dosi (1981) suggests that the leading US firms followed a more conservative mark-up pricing strategy, rather than the very aggressive long-term, loss-leadership strategy. Both agree that in any case entry barriers to new firms are now formidable, both because of R & D costs and because of other static and dynamic scale economies. In Dosi's view, the industry has now become a mature international oligopoly. The implications of costs of innovation in relation to size of firm are discussed further in chapters 5 and 6, but this chapter concludes by considering the effects of microelectronic technology on other sectors of the economy.

The microelectronic revolution

Almost every industrial and service activity is already affected to some extent by the microelectronic revolution described above, since all require information and in many activities the rapid recording, storage, processing and communication of information is the main function of the firms involved, as for example with banks and insurance companies. The speed, scale, reliability and cost of microelectronics means that the range of industrial and office processes affected by automation can now be greatly extended. Automation had already gone a long way, on the basis of earlier electro-mechanical and electronic systems, before the advent of the microprocessor and the microcomputer in the 1970s. The microprocessor both enhanced the performance of many existing electronic products and made it possible to introduce a great many new ones.

However, the fact that a new technology has many potential applications does not of course mean that all of these will occur simultaneously, or even over a short period. On the contrary, the assimilation of a major new technology into the economic and social system is a matter of decades, not of years, and may be related to the phenomenon of long cycles in the economy, as Schumpeter (1939) originally suggested (Freeman, Clark and Soete, 1982). This prolonged diffusion process is more or less inevitable, because a major new technology, such as steam power, electric power or electronics, involves a great many enabling educational, social and managerial changes, as well as clusters of technical innovations in a great variety of applications, and many scaling-up processes. This applies *a fortiori* to the international diffusion of new technological systems.

In the case of electronics, visionary scientists and engineers, such as Wiener (1949) and Diebold (1952), were essentially right thirty years ago when they foresaw the many potential applications of electronic computers throughout the economic system in both factories and offices. Where they went badly wrong was in their estimation of the time scale. Wiener failed to take account of the long time lags in building up a capital goods supply industry and a component industry on a sufficient scale to provide all the computer power, peripherals and instrumentation for this vast transformation. Even more, perhaps, he underestimated the time scale needed to educate and train millions of people in the design, redesign, operation and maintenance of a huge variety of processes incorporating the new technology. Finally, he took insufficient account of the relative costs of the new technology which was still unattractive in purely economic terms for many potential applications.

As we have seen, during the 1950s, 1960s and 1970s there was indeed a massive expansion of the supply potential of the electronic capital goods and component industries and an enormous improvement in their relative costs and reliability compared with older electro-mechanical systems. But this still does not mean that there can be a sudden introduction of the new technologies throughout the economy, for several very important reasons. First of all, there is still a skill shortage especially in those sectors which have never had any wide experience of electronics technology. Secondly, the software costs in designing entirely new applications

Table 4.7   Diffusion of microelectronic technology through the economy: An illustrative table. (This table is not intended to give precise data but to illustrate some of the main trends.)

Rate of Diffusion*	Rapid	Medium		Slow		
(Depth of impact)†	High	High	Medium	High	Medium	Low
Design and redesign of products to use microelectronic technology	Electronic capital goods; Military and space equipment; Some electronic consumer goods	Machine tools; Vehicles; Electronic consumer goods; Instruments; Some toys	Other consumer durables; Engines and motors; Other machinery	Some biomedical products	Other toys	Agriculture; Hotels and restaurants; Construction; Personal services
Process automation using microelectronic technology	Some electronic products	Machining (batch and mass) especially in vehicles, consumer durables and machinery; Printing and publishing; Design	Continuous flow processes already partly automated e.g.: • chemicals • metals • petroleum • gas • electricity	Clothing; Textiles; Food; Assembly	Building materials; Furniture; Mining and quarries	Agriculture; Hotels and restaurants; Construction; Personal services
Information systems and data processing	Specific government, business and professional systems involving heavy data storage and processing in large organisations	Financial services; Communication systems; Office systems and equipment without total electronic systems	Transport; Wholesale distribution; Public administration; Large retailers	Retail distribution; All-electronic office systems; Electronic funds transfer	Domestic households; Professional services	Agriculture; Hotels and restaurants; Construction; Personal services

*Ranging from less than 10 years (rapid) to more than 30 years (slow) for the greater part of production to be affected.
† Proportion of total product or process equipment cost.
Source: Freeman (1982).

can be extremely high even though the hardware costs have been drastically reduced. Thirdly, full-scale automation may only be possible in association with other heavy re-equipment investment. Finally, in some important service sectors it may only be possible as a result of legislative, organizational, managerial and social changes which take a long time to bring about.

A pervasive new technology is likely to find application first of all in the rapidly growing new sectors of the economy, where a lot of new investment is taking place in any case and where there is greater acceptance of innovation. Thus the main applications of microelectronics have initially been in the electronics industry itself (Table 4.7).

The second group of firms to make early and extensive use of the new technology were those where electronic subsystems represented a large part of total product cost and where the necessary skills were either already available (as sometimes in the case of scientific instruments and cash registers) or could be injected by aggressive strategies on the part of component suppliers (as in the case of calculators).

A third group of firms to use electronic technology rather quickly were those already operating large-scale flow production systems, especially those with innovative management facing expanding markets, as in the case of the chemical industry and the electric-power industry in the 1950s and 1960s. Flow processes were already partly automated using older techniques. All of these three types of firm were able to achieve very big and sustained increases in labour productivity in the post-war period through the combined advantages of the new technology and the exploitation of scale economies.

When, however, we come to consider the older and slower growing (or declining) sectors of the economy, especially those with a low endowment of qualified engineers, then the problems of diffusion are very different. Empirical diffusion studies suggest that thirty years is by no means uncommon for the time period over which revolutionary innovations are diffused through the majority of a potential adopter population and it can be much longer. These big variations in the responsiveness of different sectors of the economy to the diffusion of new technology are illustrated in Table 4.7. They account for much of the controversy about the impact of microelectronics where those involved are often talking at cross purposes because they are referring to different sectors of the economy, where different time scales apply.

Part Two

INNOVATION AND THE THEORY OF THE FIRM

In capitalist reality as distinguished from its textbook picture, it is not that kind of competition which counts (competition that is, within a rigid pattern of invariant conditions of production) but the competition from the new commodity, the new technology, the new source of supply, the new type of organization . . . competition which strikes not at the margins of the profits, and the outputs of existing firms, but at their very lives. This kind of competition is as much more effective than the other as a bombardment is in comparison with forcing a door, and so much more important that it becomes a matter of comparative indifference whether competition in the ordinary sense functions more or less promptly. . . .
Schumpeter (1942, p. 84)

The prime essential to that perfect competition which would secure in fact those results to which actual competition only 'tends' is the absence of uncertainty (in the true unmeasurable sense). Other presuppositions are mostly included in or subordinate to this, that men must *know what they are doing*, and not merely guess more or less accurately.
Knight (1965, p. 20)

The bourgeoisie cannot exist without constantly revolutionizing the means of production, and thereby the relations of production, and with them the whole relations of society. Conservation of the old modes of production in unaltered form was, on the contrary, the first condition of existence for all earlier industrial classes. Constant revolutionizing of production, uninterrupted disturbance of all social conditions, everlasting uncertainty and agitation distinguish the bourgeois epoch from all earlier ones. . . . The bourgeoisie has through its exploitation of the world market given a cosmopolitan character to production and consumption in every country. To the great chagrin of reactionaries, it has drawn from under the feet of industry the national ground on which it stood. All old-established national industries have been destroyed or are daily being destroyed. They are dislodged by new industries, whose introduction becomes a life and death question for all civilized nations. . . .
Marx and Engels (1848)

5 Success and Failure in Industrial Innovation

We now consider some tentative generalizations about the technical innovation process in the industries described, and discuss how far it is possible to test the validity of such generalizations, and to relate them to other industries and the economy as a whole.

Jewkes and his colleagues (1958) have argued that the ninteenth-century links between science and invention were much greater than is commonly assumed. Certainly, the classical economists were well aware of the connection between scientific advances and technical progress in industry, in the eighteenth and early nineteenth centuries. The quotation from Adam Smith (1776), with which this book begins, illustrates this point. Nevertheless, the evidence of the previous three chapters suggests that there were profound changes in the degree of intimacy and the nature of the relationship between science and industry.

The contact in the eighteenth and early nineteenth centuries was spasmodic and unsystematic. Few firms had scientists working for them, although it is true that scientists were consulted occasionally about industrial problems. There were in any case very few scientists. Many inventions and innovations in the textile, metal-working and railway industries owed little or nothing to scientific research. They were based far more on the practical experience of engineers and craftsmen. Scientist–entrepreneurs and inventor–entrepreneurs did play a very important part in the chemical industry and in some branches of mechanical engineering (Musson and Robinson, 1969). But with the growth of the dyestuffs industry and the electrical engineering industry in the second half of the nineteenth century there was a change in the pattern of relationships, which became more clearly established in the twentieth-century industries of electronics, synthetic materials and flow-process plant. Not only were the products and processes of these industries originally based almost entirely on scientific discóveries and theories, but even the day-to-day improvement and modification of the products and processes depended to an increasing extent on an understanding of scientific principles and on laboratory experiments.

As already explained in chapter 1, this does not imply the acceptance of a 'linear' model of R & D with a simple one-way flow of ideas from basic science through applied research to development and commercial innovation. On the contrary, there has always been and there remains in the modern science-related industries a strong reciprocal interaction between all these activities, and in particular a powerful influence of technology upon science. Gazis (1979) has given some recent examples of this interaction in the case of IBM's research laboratories. However, the effectiveness of this two-way movement of ideas depends on the ability of both communities to communicate with each other.

The new style of innovation in the industries which we are considering was characterized by professional R & D departments within the firm, employment of qualified scientists as well as engineers with scientific training, both in research and in other technical functions in the firm, regular contact with universities and other centres of fundamental research, and acceptance of science-based technical change as a way of life for the firm. Some of the firms we have considered had very strong scientific and technical resources, such as ICI, BASF, Du Pont, Hoechst, RCA, Marconi, Telefunken and Bell. An extreme case was the development of nuclear weapons and atomic energy.

Almost all of the major innovations we have considered were the result of professional R & D activity, often over long periods (PVC, nylon, polyethylene, terylene, synthesis of ammonia, hydrogenation, catalytic cracking, nuclear power, television, radar, semi-conductors . . .). Even where individual inventor–entrepreneurs played the key role in the innovative process (at least in the early stages) such individuals were usually scientists or engineers who had the facilities and resources to conduct sustained research and development work (Baekeland, Fessenden, Eckert, Houdry, Dubbs, Marconi, Armstrong, Zuse). Some of them used university or government laboratories to do their work, while others had private means.

Frequently university scientists or inventors worked closely as consultants with the corporate R & D departments of the innovating firms (Ziegler, Natta, Haber, Fleming, Michels, Staudinger, Von Neumann). In other cases special war-time programmes led to the recruitment of outstanding university scientists to work on government-sponsored innovations (the atomic bomb and radar). Intimate links with basic research through one means or another were normal for R & D in these industries and their technology is science-based in the sense that it could not have been developed at all without a foundation in theoretical principles. This corpus of knowledge (macromolecular chemistry, physical chemistry, nuclear physics and electronics) could never have emerged from casual observation, from craft skills or from trial and error in existing production systems, as was the case with many earlier technologies.

The rise of these new science-related technologies has had major economic and social repercussions over and above the growth of professional industrial R & D. It changed not only the development procedures, but also the production engineering, the sales methods, the industrial training and the management techniques. Quite often the majority of employees in firms in the new industries were not employed in production or handling of goods at all, but in generating, processing and distributing information. In the extreme cases the computer software or process plant design and consultancy firm may employ hundreds of people but have no physical output other than paper or computer print-out. But even in quite 'normal' electronic or chemical firms, the combined employment in research, development, design, training, technical services, patents, marketing, market research and management may be greater than in 'production'. The complexity of the technical information involved and of the data processing means that specialized information storage, handling and retrieval systems are increasingly necessary. This proliferation of 'nonproduction' occupations is often treated as a form of Parkinson's Law or conspicuous waste. Even a scientist–inventor such as Gabor (1964) treated it as unnecessary in economic terms (although perhaps desirable on social grounds). No doubt Parkinson's Law does operate, and no doubt manpower savings can be made in some of these occupations (as they can in production). But it is essential to this analysis that the major part of this growth is due to the changes in technology, and to the new forms of competition which this has brought about.

So far we have discussed the new industries mainly in terms of the scientific basis of their new products and their manufacturing technologies, but it is impossible to disregard the pull of the market as an essential complementary force in their origins and growth. In many cases the demand from the market side was urgent and specific.

The strength of the German demand for 'ersatz' materials to substitute for natural materials in two world wars spurred on the intense R & D efforts of IG Farben and other chemical firms. The strength of the military/space demand in the American post-war economy stimulated the flow of innovations based on Bell's scientific breakthrough in semiconductors and the early generations of computers. The urgency of British wartime needs spurred the successful development of radar of all kinds while the German Government sponsored the development of FM networks, as well as radar.

Conversely, the absence of a strong market demand for some time retarded the development of synthetic rubber in the US, the growth of the European semi-conductor industry, the development of radar in the US before 1940, or of colour television in Europe after the war.

This does not mean that only wartime needs and government markets can provide sufficient stimulus for innovations, although they were obviously important historically. A strong demand from firms for cost-reducing innovations in the chemical and other process industries is virtually assured, because of their strong interest in lower costs of producing standard products and their technical competence. The demand for process innovations is related to the size of the relevant industry and here again the American oil industry provided a key element of market pull for the innovative efforts of the process-design organizations. The market demand may come from private firms, from government or from domestic consumers, but in its absence, however good the flow of inventions, they cannot be converted into innovations.

Innovation is essentially a two-sided or coupling activity. It has been compared by Schmookler (1966) to the blades of a pair of scissors, although he himself concentrated almost entirely on one blade. On the one hand, it involves the recognition of a need or more precisely, in economic terms, a potential market for a new product or process. On the other hand, it involves technical knowledge, which may be generally available, but may also often include new scientific and technological information, the result of original research activity. Experimental development and design, trial production and marketing involve a process of 'matching' the technical possibilities and the market. The professionalization of industrial R & D represents an institutional response to the complex problem of organizing this 'matching', but it remains a groping, searching, uncertain process.

In the literature of innovation, there are attempts to build a theory predominantly on one or other of these two aspects. Some scientists have stressed very strongly the element of original research and invention and have tended to neglect or belittle the market. Economists have often stressed most strongly the demand side: 'necessity is the mother of invention'. These one-sided approaches may be designated briefly as 'science-push' theories of innovation and 'demand-pull' theories of innovation (Langrish *et al.*, 1972). Like the analogous theories of inflation, they may be complementary and not mutually exclusive.

In a powerful critique of demand-pull theories of innovation, Mowery and Rosenberg (1979) have pointed to the inconsistent use of the concept of 'demand' in this literature and insist that the results of recent empirical surveys of innovation cannot legitimately be used (although they often have been) to support one-sided market-pull theories. The example of the electronic computer cited in chapter 4 is a good example of their point that the 'market' cannot evaluate a revolutionary new product of which it has no knowledge.

It is not difficult to cite instances which appear to give support to either theory. There are many examples of technical innovation, such as the atomic absorption spectrometer, where it was the scientists who envisaged the applications without any very clear-cut demand from customers in the early stages. Going even further, advocates of 'science-push' tend to cite examples such as the laser or nuclear power, where neither the potential customers nor even the scientists doing the original work ever envisaged the ultimate applications or even denied the possibility, as in the case of Rutherford. Advocates of 'demand-pull' on the other hand tend to cite examples such as synthetic rubber or cracking processes or photo-destruction of plastics, where a clearly recognized need supposedly led to the necessary inventions and innovations (see, for example, *Sunday Times*, 1970).

Whilst there are instances in which one or the other may appear to predominate, the evidence of the innovations considered here points to the conclusion that any satisfactory theory must simultaneously take into account *both* elements. Since technical innovation is defined by economists as the first *commercial* application or production of a new process or product, it follows that the crucial contribution of the entrepreneuer is to *link* the novel ideas and the market. At one extreme there may be cases where the only novelty lies in the idea for a new market for an existing product;[1] at the other extreme, there may be cases where a new scientific discovery automatically commands a market without any further adaptation or development. The vast majority of innovations lie somewhere in between these two extremes, and involve some imaginative combination of new technical possibilities and market possibilities. Necessity may be the mother of invention, but procreation still requires a partner.

Almost any of the innovations which have been discussed could be cited in support of this proposition. Marconi succeeded as an innovator in wireless communication because he combined the necessary technical knowledge with an appreciation of some of the potential commercial applications of radio. The Haber–Bosch process for synthetic ammonia involved both difficult and dangerous experimental work on a high pressure process and the development of a major artificial fertilizer market, stimulated by fears of war and shortage of natural materials. Despite their early complete underestimation of the market, IBM were the most successful firm in the world computer industry because they combined the capacity to design and develop new models of computers, with a deep knowledge of the market and a strong selling organization. Firms such as General Electric and RCA with similar or greater scientific and technical strength, but much less market knowledge and market power in this field, in the end had to withdraw.

We may indeed advance the proposition that 'one-sided' innovations are much less likely to succeed. The enthusiastic scientist–inventor or engineer who neglects the specific requirements of the potential market or the costs of his product in relation to the market is likely to fail as an innovator. This occurred with EMI and AEI in computers and with several British firms in radar, despite their technical accomplishments and strong R & D organizations. Professionalization of industrial R & D means that there is now often an internal 'pressure group' which may push 'technologically sweet' ideas without sufficient regard to the potential market, sales organization or costs.

On the other hand, the entrepreneur or inventor–entrepreneur who lacks the necessary scientific competence to develop a satisfactory product or process will fail as an innovator, however good his appreciation of the potential market or his selling. This was the fate of Parkes with his plastic comb and of Baird with television.

[1] Whilst this may be described as innovation, it cannot be legitimately described as *technical* innovation. Non-technical organizational innovations are extremely important and often associated with technical innovations but they are not discussed here.

The failures may nevertheless contribute to the ultimate success of an innovation, even though the individual efforts fail. The social mechanism of innovation is one of survival of the fittest. The possibility of failure for the individual firms which attempt to innovate arises *both* from the technical uncertainty inherent in innovation *and* from the possibility of misjudging the future market and the competition. The notion of 'perfect' knowledge of the technology or of the market is utterly remote from the reality of innovation, as is the notion of 'equilibrium'.

The fascination of innovation lies in the fact that both the market and the technology are continually changing. Consequently, there is a kaleidoscopic succession of new possible combinations emerging. What is technically impossible today may be possible next year because of scientific advances in apparently unrelated fields. Although he developed the concept mainly in relation to invention rather than technical innovation, Usher's 'Gestalt' theory probably comes close to representing the imaginative process of 'matching' ideas. What cannot be sold now may be urgently needed by future generations. An unexpected turn of events may give new life to long-forgotten speculations or make today's successful chemical process as dead as the dodo. Patents for a float glass process and for radar were taken out before 1914. The stone that the builders rejected is the corner-stone of the arch. The production of polyethylene was nearly suspended after the Second World War because the peacetime markets were thought to be too small. IG Farben offered to sell their synthetic-rubber patents to the natural-rubber cartel because they thought the synthetic product would not be able to compete in peacetime in price or quality. The early computer manufacturers expected that the market would be confined to government and scientific users. A century after early experiments steam-driven road vehicles were again being seriously investigated by major automobile manufacturers. The apparently random, accidental and arbitrary character of the innovative process arises from the extreme complexity of the interfaces between advancing science, technology and a changing market. The firms which attempt to operate at these interfaces are as much the victims of the process as its conscious manipulators. Innovation works as a social process but often at the expense of the innovators. The implications of this high degree of uncertainty are discussed in chapter 7.

These considerations lead to three conclusions of fundamental importance. First, since the advance of scientific research is constantly throwing up new discoveries and opening up new technical possibilities, a firm which is able to monitor this advancing frontier by one means or another may be one of the first to realise a new possibility. Strong in-house R & D may enable it to convert this knowledge into a competitive advantage. Secondly, a firm which is closely in touch with the requirements of its customers may recognize potential markets for such novel ideas or identify sources of consumer dissatisfaction, which lead to the design of new or improved products or processes. In either case, of course, they may be overtaken by faster-moving or more efficient competitors or by an unexpected twist of events, whether in the technology or in the market. Thirdly, the test of successful entrepreneurship and good management is the capacity to link together these technical and market possibilities, by combining the two flows of information.

Innovation is a coupling process and the coupling first takes place in the minds of imaginative people. An idea 'gels' or 'clicks' somewhere at the ever-changing interfaces between science, technology and the market. For the moment this begs the question of 'creativity' in generating the inventive idea, except to note that almost all theories of discovery and creativity stress the concept of imaginative association or combination of ideas previously regarded as separate. But once the idea has 'clicked' in the mind of an inventor or entrepreneur, there is still a long way to go before it becomes a successful innovation, in our sense of the term. Rayon was 'invented' 200 years before it was innovated, the computer at least a century before, and aeroplanes even earlier.

The 'coupling' process is not merely one of matching or associating ideas in the original first flash; it is far more a continuous creative dialogue during the whole of the experimental development work and introduction of the new product or process. The one-man inventor–entrepreneur like Marconi or Baekeland may very much simplify this process in the early stages of a new innovating firm, but in the later stages and in any established firm the 'coupling' process involves linking and coordinating different sections, departments and individuals. The communications within the firm and between the firm and its prospective customers are a critical element in its success or failure. As we have seen, in many cases the original idea may take years or even decades to develop, and during this time it continually takes on new forms as the technology develops and the market changes or competitors react. Consequently, the quality of entrepreneurship and good communications are fundamental to the success of technical innovations.

Summing up this discussion and the evidence in Part One, we might conclude that among the characteristics of successful innovating firms in the industries considered were:

1 Strong in-house professional R & D
2 Performance of basic research or close connections with those conducting such research.
3 The use of patents to gain protection and to bargain with competitors.
4 Large enough size to finance fairly heavy R & D expenditures over long periods.
5 Shorter lead-times than competitors.
6 Readiness to take high risks.
7 Early and imaginative identification of a potential market.
8 Careful attention to the potential market and substantial efforts to involve, educate and assist users.
9 Entrepreneurship strong enough effectively to coordinate R & D, production and marketing.
10 Good communciations with the outside scientific world as well as with customers.

We might hypothesize that these are the essential conditions for successful technical innovation.

Up to a point, such tentative generalizations about the characteristics of innovation may be tested by analysing and comparing case studies of a large number of innovations. One difficulty about such case studies (many of which have been cited in Part One) is that we do not know how far they are representative of the innovative process. Indeed, much of the literature on industrial innovation falls into two categories: scattered case histories lacking comparability of coverage or theoretical analysis lacking systematic empirical foundations.

As a result of this, there are many plausible, half-tested hypotheses and many interesting conjectures in innovation theory, but insufficient firm evidence to refute or support them. The historical account in Part One suggests interesting conclusions, but it is difficult to find ways to substantiate them, or to assess their relative importance. Yet such systematic testing of generalizations and hypotheses is essential to advance our understanding.

The remainder of this chapter is therefore devoted to the description of a project which was deliberately designed to test such generalizations about innovation. The project was called SAPPHO and it was carried out at the Science Policy Research Unit during the 1970s. The original project was designed in 1968 by R. C. Curnow, but later stages of the work were led by R. Rothwell.

Project SAPPHO

The basic idea of the project was to attempt to substantiate (or refute) generalizations about technical innovation, by the systematic comparison of 'pairs' of successful and unsuccessful attempts to innovate in each branch of industry in turn. This method of course rests on the observation that competitive technical innovation is a fairly general characteristic of many branches of industry in industrialized capitalist societies.

Since the introduction of a new product or a new process in any branch of industry may render older products and processes obsolete or uneconomic, firms which wish to survive and grow must be capable of adapting their technologically-based strategy to this competition. This does not necessarily mean that every firm has to be research-minded or to innovate itself. Various alternative strategies are possible for the firm even in an industry subject to rapid technical change. Some firms may even prefer to disappear rather than to innovate. These alternatives are considered more systematically in chapter 8. For the time being we are concerned with those firms who *do* attempt to innovate, whether in products or processes. Some of these firms may attempt to be the first to introduce a new product or process hoping thereby to gain a technological lead and temporary monopoly profits. This strategy is sometimes designated as 'offensive' innovation. Others may act only 'defensively' in response to innovations introduced by competitors. In either case the firm will need some capacity to develop and launch new products or processes (even if under a licensing agreement or by simple imitation). Frequently a firm which attempts to be first may not succeed, and multi-product firms may be 'offensive' in some fields and 'defensive' in others. But in the long run their survival and growth will depend on whether they succeed or fail in their innovations, whether 'offensive' or 'defensive'.

The first stage of project SAPPHO, which is summarized here, was a study of fifty-eight attempted innovations in chemicals and scientific instruments (listed in Table 5.1). Those in the chemical industry were mainly process innovations, whilst the instrument innovations were all product innovations. The instruments were mainly electronic and the chemical processes related mainly to intermediates derived from petroleum (SPRU, 1972). In later work, additional pairs of innovations were studied in these same industries and then, using a somewhat different methodology, paired comparisons were made in various sectors of the mechanical engineering industry (Rothwell *et al.*, 1974; Rothwell, 1976).

By 'pairing' attempted innovations it was hoped to discriminate between the respective characteristics of failure and success. The technique had of course been widely used in the natural sciences, especially in biology (McKay and Bernal, 1966). When the two halves of a pair differ with respect to a particular characteristic or set of characteristics, this indicates a possible 'explanation' of innovative success or failure. Where there is a significant and repeated variation between the pattern of 'success' and 'failure', across a large number of pairs, this provides systematic evidence for the validity of particular hypotheses or groups of hypotheses.

Table 5.1   List of SAPPHO pairs

Scientific instruments	Chemicals
Amlec eddy-current crack detector	Accelerated freeze-drying of food (solid)
Atomic absorption spectrometer	Acetic acid
Digital voltmeters	Acetylene from natural gas
Electromagnectic blood-flow meter	Acrylonitrile I
Electronic checkweighing I	Acrylonitrile II
Electronic checkweighing II	Ammonia synthesis
Foreign-bodies-in-bottles detector	Caprolactam I
Milk analysers	Caprolactam II
Optical character recognition	Ductile titanium
Roundness measurement	Extraction of aromatics
Scanning electron microscope	Extraction of n paraffins
X-ray microanalyser	Hydrogenation of benzene to cyclohexane
	Methanol
	Oxidation of cyclohexane
	Phenol
	Steam naphtha reforming
	Urea manufacture

Such explanations as appear to have a significant statistical foundation may then be tested again on a new sample of innovations. In this way a structured and tested foundation for theoretical work may be built up.

It was expected that the success and failure halves of a pair would resemble each other fairly closely in many respects, and this proved to be the case. It could be assumed from previous experience that firms attempting to develop a particular new process or product would often have many characteristics in common. The analysis of similarity is complementary to the analysis of divergence for two reasons. First, it enables the identification of some characteristics which are shared by all firms attempting innovation in particular industries. These may be necessary conditions for entry into the race, and may be regarded as such unless other 'success' cases can be found which disprove such tentative generalizations. But secondly, and more important, they enable us to focus attention on those characteristics in which the pattern *does* diverge between success and failure. In future research it will be possible to concentrate in greater depth on these significant differences through a process of elimination of unnecessary hypotheses.

The pairs were not 'identical twins'. Their similarity was defined in terms of their market, not necessarily in terms of their technology. For example, two firms might both be seeking a new, cheaper and better way to produce phenol or urea. They might adopt somewhat different technical solutions. It was an assumption of this project that this very choice constituted part of the 'success', and the 'wrong' choice part of the 'failure'. In a few cases the resemblance was very close, as where several licensees shared access to the same basic technical knowledge. But even here the design varied when two manufacturers attempted to satisfy the same demand. Success depends partly on developing the 'right' design, having regard to the available scientific and technical knowledge *and* to the potential uses of the innovation.

Concentration on *innovation* rather than invention has many consequences in terms of method. The most important of these is that the marketing aspects of the process assume much greater importance, whilst the role of that individual, usually described as *the* inventor, recedes into a wider social context. Our comparisons did not include those numerous experiments by inventors and would-be innovators which are discarded or shelved long before they reach the point of commercial introduction. Such studies are undoubtedly of interest in the management of R & D, but the focus here was on the wider problem of the management of innovation. The 'failures' were products or processes which were brought to the point of commercial introduction, and usually were in fact on the market for some years. Attention was therefore concentrated both on the various stages of development work and also on the preparations for production and sale and the experience of marketing the innovation.

Since the project was concerned with technical innovation in industry, the criterion of success was a commercial one. A 'failure' is an attempted innovation which failed to establish a worthwhile market and/or make any profit, even if it 'worked' in a technical sense. A 'success' is an innovation which attained significant market penetration and/or made a profit. This chapter analyses twenty-nine 'successes' and twenty-nine 'failures'. Often a failure was clear-cut, e.g. a firm went bankrupt, or closed a plant down, or withdrew a product or failed to sell it, while success was not always so self-evident. A product might achieve a world-wide market, but take a long time to show a profit. One case (corfam) which was originally expected to be a 'success' was withdrawn from the market on these grounds (chapter 3). Even with failures it was not always simple to make an assessment. There were varying shades of grey between the extremes of success and failure. The project deliberately tried to investigate the fairly clear-cut 'black and white' cases of failure and success. In two cases in the chemical industry, and one in instruments, it proved to be feasible to complete two pairs, as there were several commercial successes and several less successful attempts in each case.

Earlier work on the literature survey had shown that there were many possible explanations of success and failure. The project was therefore designed to test a large number of single hypotheses and simultaneously to test a large number of possible combinations of factors. The aim was to identify a characteristic *pattern* of failure or success. Altogether about two hundred measurements of each case of 'success' or 'failure' were attempted. Some of the measures were comparative, some absolute. Thus, for example, it was possible to test the hypothesis that large size is generally advantageous for innovation, both by testing in how many pairs the smaller of the two firms failed, and by checking what proportion of firms with fewer than 500 employees succeeded (or 100, 1,000 or 10,000 employees).

But most of the measures were comparisons between the success and failure halves of the pair, enabling statements to be made such as 'successful attempts were characterized by greater . . . or less . . . or smaller . . . or shorter . . . or more . . .' than attempts which failed, but the aim was to link all those comparisons together to derive a pattern of success. The main hypotheses which the project attempted to test related to various measures of size (employment, R & D department and project team); measures of market research, publicity, education of users, involvement of users; modification of the innovation and checks on its progress at various stages; the role of engineers and scientists and of various key individuals, their previous experience, education and background;[2] the management control and planning system in the firm; the communication network with

[2] Four key 'roles' in the conduct of innovation were defined as follows:

(a) *'Technical innovator'*. The individual who made the major contributions on the technical side to the development and/or design of the innovation. He would normally, but not

the outside world; degree of dependence on outside technology and familiarity of the firm with the innovation; effectiveness and methods of organizing R & D work, patent policy, competitive pressures, speed in development work and date of commercial launching.

The results of the analysis may be classified under three headings:

1 Factors which were common to almost all attempts to innovate, whether successful or not.
2 Factors which varied between innovative attempts, but in which the variation was not systematically related to success or failure.
3 Measures which discriminated between success and failure.

1 Resemblance between pairs

Taking the first twenty-nine pairs, involving fifty-eight attempts to innovate, there were many resemblances between both halves of the pairs (Table 5.2). Almost all attempts in these two industries took place within a formal R & D structure which was used to develop the innovation. This confirms the professionalization of R & D, with the reservations made on p. 128. Only in the instrument industry were there cases of attempted innovation without such a structure. Most of these were designs for a new product brought from an outside environment.

Since almost all of the firms involved in attempts to innovate had this formal R & D structure, it might be expected that critical differences would exist in the way in which such departments were organized, R & D was planned, projects evaluated, or incentives provided for engineers and scientists. A great deal of the management and sociological literature has concentrated on these aspects of the efficiency of industrial research.

The inquiry did not uncover systematic differences of this kind with respect to R & D organization or incentives. As with previous empirical studies it was found that many supposedly 'best practice' techniques in long-term planning and project assessment were honoured more in the breach than in the observance. But the successful innovators differed only a little from the failures in this respect. In the chemical industry, although not in instruments, there was some evidence

    necessarily, be a member of the innovating organization. He would sometimes, but not always, be the 'inventor' of the new product or process.

(b) *'Business innovator'*. That individual who was actually responsible within the management structure for the overall progress of this project. He might sometimes be the technical director or the research director. He might be the same man as the 'technical innovator'. He could be the sales director, or chief engineer. Occasionally, especially in smaller firms, he could be the chief executive for the organization as a whole.

(c) *'Chief executive'*. The individual who was formally the head of the executive structure of the innovating organization, usually but not necessarily with the job title of 'managing director'. In every case there was an identifiable 'chief executive', and almost always an identifiable 'business innovator', but quite often there was no identifiable 'technical innovator'. No attempt was made to force individuals to assume these roles if they were not readily identifiable, since one of the objects of the inquiry was to assess the contribution of outstanding individuals.

(d) *'Product champion'*. Any individual who made a decisive contribution to the innovation by actively and enthusiastically promoting its progress through critical stages. He might sometimes be the same individual as the technical innovator, or 'chief executive'. Although these roles have been recognised in much of the earlier innovation literature, they are not always identifiable from formal titles used in firms. The job title might vary a good deal, but it was the role which was important.

that better management and planning techniques were associated more frequently with success. Some of the difficulties inherent in R & D forecasting and project evaluation are discussed in chapter 7. There was no evidence that successful innovators expected rewards or penalties differing from the less successful nor was there any evidence of unusual incentive schemes for R & D personnel, or greater freedom in successful cases.

One possible explanation of more successful attempts to innovate might lie in patent priority, but again it was not possible to identify differences here. Almost all innovators, both successful and unsuccessful, took out patents and regarded them as important. But the failures did not attribute their lack of success to the patent position of their rivals, except in one case. These results confirm the evidence of the historical account that innovating firms usually take trouble over patents because of their importance as bargaining counters, but that patents do not necessarily prevent any competitive developments.

Nor was it found that successful innovators differed from unsuccessful in the way in which they organized their project teams. One hypothesis had suggested that the less successful attempts might be characterized by departmental organization on disciplinary lines. But this was not the case. Where firms had a large R & D organization, they sometimes had laboratories working on conventional sub-disciplinary lines, but this did not really affect the project *development* team which was set up in a similar way in both successes and failures.

Another hypothesis for which no supporting evidence was found was the view that business or technical innovators might be less well qualified academically in unsuccessful attempts (or better qualified). There were important differences between business innovators which *did* distinguish between success and failure, but this was not one of them. Most of the business innovators, and almost all the technical innovators, were qualified scientists or engineers in *both* halves of the pair. There was a slight tendency for the Ph.Ds to be the more successful in chemicals. Obviously, in these two industries amateurs are rarely chosen to manage innovations. In the two cases when accountants were the business innovators, both failed, but this would be too small a number on which to construct any general theory. It would probably not be possible to find a sufficient number of cases in these industries where innovators were *not* technically qualified, to test any hypotheses relating to the supposed merits or de-merits of amateurism. The difficulty of finding such cases is, however, further evidence of the professionalization of the innovative process.

2 Variation unrelated to success or failure

Many other measures *did* show considerable variation between attempts to innovate, but the variance was not closely related to success. Among them were measures relating to size of firm, size of R & D department and numbers of qualified engineers and scientists in R & D. These results need considerable care in interpretation. There was no strong systematic evidence that larger or smaller firms or R & D departments were more or less successful. For example, of the cases involving firms employing more than 10,000 people, six were successes and seven were failures. Where large firms were in competition with smaller firms there was a tendency for them to be more successful, but it was by no means clearcut. At first sight this finding is perhaps at variance with some of the evidence from the historical descriptive account in Part One, which suggested that the heavy costs and long time scale of many innovations would give an advantage to large firms.

Table 5.2 Part 1   Some measures which did not differentiate between success and failure

Question	Chemicals			Instruments			Both industries			Binomial test
	$S > F$	$S = F$	$S < F^a$	$S > F$	$S = F$	$S < F^a$	$S > F$	$S = F$	$S < F^a$	
Was the innovation more or less radical for the firms concerned?	5	7	5	4	3	5	9	10	10	0.5
At what level was the decision to proceed with the innovation made?	2	11	4	1	10	1	3	21	5	0.363
Was a time limit set?	4	12	1	1	8	3	5	20	4	0.5
Were patents taken out for this innovation by the organization?	—	17	—	3	8	1	3	25	1	0.313
Did one organization accept the innovation as being more in its natural business than the other?	5	8	4	6	1	5	11	9	9	0.412
Did one organization have a more serious approach to planning than the other?	6	7	4	2	7	3	8	14	7	0.5
Was there a systematic and periodically reconsidered R & D programme?	6	7	4	—	10	2	6	17	6	0.613
What was the company's publishing policy?	5	7	5	1	11	—	6	18	5	0.623
Were there any incentive schemes to encourage innovative effort?	2	15	—	—	12	—	2	27	—	0.25
What outcome was the project expected to have on the careers of members of the project team in the event of success?	3	11	3	1	11	—	4	22	3	0.5
Was the innovation part of a general marketing policy?	5	8	4	2	9	1	7	17	5	0.387

Question										
What was the degree of coupling with the outside scientific and technological community in general?	2	12	3	2	8	2	4	20	5	0.5
Would the firm have recruited more QSEs if it could have done so at the time of the innovation?	–	17	–	1	10	1	1	27	1	0.75
In each case, when was the decision to innovate formalized on paper?	5	5	7	1	10	1	6	15	8	0.395
How many months elapsed from prototype or pilot plant to first commercial sale?	7	3	7	5	3	4	12	6	11	0.5
Was there a formal R & D department in the organization?	1	16	–	2	8	2	3	24	2	0.5
What was the scale of growth of the organization up to the time of marketing (measured by annual growth of turnover in the five years prior to the marketing of the innovation)?	1	13	3	4	4	4	5	17	7	0.387
How many years did the business innovator spend in the educational system?	5	7	5	2	5	5	7	12	10	0.315
Was the R & D department regarded as a profit centre?	6	8	3	1	10	1	7	18	4	0.274
Was there any need to find or use new materials?	–	17[b]	–	–	10	1	–	27	1	0.5

[a] S > F   Success more than failure, greater than failure, etc., *or* in success but not in failure.
  S = F   No measurable difference between success and failure.
  S < F   Success less than failure, smaller than failure, etc., *or* in failure but not in success.
[b] Data not available in one case.
– nil.

Table 5.2 Part 2   Some measures which differentiate between success and failure

Question	Chemicals			Instruments			Both industries			Binomial test
	$S > F$	$S = F$	$S < F$[a]	$S > F$	$S = F$	$S < F$	$S > F$	$S = F$	$S < F$	
Was the innovation more or less radical for world technology?	10	6	1	2	9	1	12	15	2	0.0065
How deliberately was the innovation sought, comparatively?	7	8	2	6	6	—	13	14	2	0.0037
Was there opposition to the project within the total organization on commercial grounds?	1	9	7	1	7	4	2	16	11	0.0112
Was more use made of development engineers in planning and costing for production in one case than in the other?	5	9	3	4	8	—	9	17	3	0.073
Did one organization have a more satisfactory communication network than the other externally?	5	10	2	5	7	—	10	17	2	0.0193
Was the R & D chief more senior by accepted status in one case than the other?	9	5	3	2	8	2	11	13	5	0.105
Was the sales effort a major factor in the success or failure of the innovation?	7	10	—	9	3	—	16	13	—	0.000015
Were any modifications introduced after commercial sales as a result of user experience?	1	8	8	2	6	4	3	14	12	0.0176
Were there any after-sales problems?	—	4	13	1	2	9	1	6	22	0.000005
Were any steps taken to educate users?	8	9	—	6	5	1	14	14	1	0.00049
If new tools or equipment were needed for commercial production, were any ordered before the decision to launch full-scale production?	8	7	2	2	10	—	10	17	2	0.227
What was the degree of coupling with the outside scientific and technological community in the specialized field involved?	8	9	—	5	6	1	13	15	1	0.00092

Question										p
How much attention was given to publicity and advertising?	6	10	1	4	7	1	10	17	2	0.0193
Did the innovation have to be adapted by users?	—	10	7	—	7	5	—	17	12	0.00024
Were there unexpected production adjustments?	1	7	9	1	7	4	2	14	13	0.00636
Did any 'bugs' have to be dealt with in the early production stage?	1	6	10	1	5	6	2	11	16	0.0049
Was any systematic forecasting by the marketing (or sales) department involved in the decision to add the innovation to product lines or to existing processes?	5	7	5	6	5	1	11	12	6	0.166
Were user needs more fully understood by the innovators in one case than in the other?	15	2	—	9	3	—	24	5	—	0.0000001
Did the business innovator have a more diverse experience in one case than in the other?	8	8	1	8	2	2	16	10	3	0.00377
Did the business innovator have a higher status in one case than in the other?	8	8	1	5	4	3	13	12	4	0.0245
Did the business innovator have more or less authority (power) in one case than in the other?	9	7	1	6	4	2	15	11	3	0.000656
To what extent was dependence on outside technology a help or a hindrance in production?	10	6	1	6	4	2	16	10	3	0.00221
How large a team was put to work on the innovation at the beginning of the project?	12	2	3	4	4	4	16	6	7	0.0466
How large a team was put to work on the innovation at the peak of the project?	9	4	4	7	4	1	16	8	5	0.0133
How many years had the business innovator spent in industry?	9	7	1	3	4	5	12	11	6	0.119
Had the business innovator had any overseas experience?	3	14	—	5	6	1	8	20	1	0.0352

Table 5.2 Part 2 *Cont.*

Question	Chemicals			Instruments			Both industries			Binomial test
	$S > F$	$S = F$	$S < F$[a]	$S > F$	$S = F$	$S < F$	$S > F$	$S = F$	$S < F$	
Did the business innovator have a greater degree of management responsibility in one case than in the other?	10	7	–	4	5	3	14	12	3	0.00636

[a] $S > F$ Success more than failure, greater than failure, etc., *or* in success but not in failure.
$S = F$ No measurable difference between success and failure.
$S < F$ Success less than failure, smaller than failure, etc., *or* in failure but not in success.
Source: Science Policy Research Unit (1971).
– nil.

However, this result should not be interpreted as implying that size of firm is completely irrelevant in relation to innovation in these two industries. *Comparative* size measured within a 'pair' did not differentiate between success and failure clearly but in chemicals only four out of thirty-four attempts were made by firms employing fewer than 1000. Clearly size is relevant to the type of innovations which are attempted at all, and inter-industry differences are very important. The next chapter is devoted entirely to a critical discussion of this problem of size in relation to innovation.

No relationship was found between success and the number of scientists and engineers on the main board of the innovating company, although this proportion varied considerably. However, in almost all cases there were *some* engineers or scientists on the main board, and it may be that this is the critical threshold factor since the innovation process requires a *combination* of technical, financial, marketing and management skills.

Perhaps surprisingly, for those who believe in the amenability of innovation to planning techniques, no relationship was found between success and the capacity to set and fulfil target dates for particular stages of the project plan, nor in the general approach to planning of the innovators. This finding too needs considerable care in interpretation and is discussed more fully in chapter 7.

Contrary to some theories, there was no association between failure and the attempt to innovate in areas unfamiliar to the firm. Where firms differed significantly in their familiarity with the field, the outcome was evenly distributed between success and failure.

Another set of measures which did not discriminate between success and failure related to the growth rate of the firm and its competitive environment. There were of course variations between firms in the growth which they had experienced before the innovation, and in the competitive pressures to which they were subject. But these differences apparently did not affect their degree of success in attempting to innovate. Again, it is important not to over-state this finding. This does not mean that competitive pressures or declining growth may not be important in stimulating attempts to innovate, only that they do not ensure success.

A rather surprising finding was that development lead-time was not strongly correlated with success. It had been expected that the more successful innovators would be those who found ways of shortening the development phase and telescoping the stages from prototype or pilot plant to commercial launch. But support for this hypothesis came only at the earlier stage of applied research. In the chemical industry successful firms were quicker to get through this early stage. The absence of any evidence of a shorter *development* stage associated with success provides support for those who have argued that hardware development is a gestation process akin to animal reproduction in that it cannot easily be artificially shortened (Burke, 1970). It may also indicate that successful firms take more trouble at the development stage to get rid of all the bugs, so that later stages are trouble-free. There was considerable indirect support for this interpretation from those measures which did discriminate between success and failure.

3 The pattern of success

Of the two hundred measures attempted, only a small number differentiated clearly between success and failure, and these varied a little between the two industries. The principal measures are shown in Table 5.2.

Those which came through most strongly were directly related to marketing. In some cases they might be regarded as obvious, but the case studies showed that even the most obvious requirements were sometimes ignored. Successful attempts were distinguished frequently from failure by greater attention to the education of users, to publicity, to market forecasting and selling (particularly in the case of instruments where it was most relevant) and to the understanding of user requirements.

The single measure which discriminated most clearly between success and failure was 'user-needs understood'. This should not be interpreted as simply, or even mainly, an indicator of efficient market research. It reflects just as much on R & D and design as well as on the management of the innovation. The product or process had to be designed, developed and freed of bugs to meet the specific requirements of the future users, so that 'understanding' of the market had to be present at a very early stage. The work of von Hippel (1976 and 1978) on 'customer-active' paradigms in new product development, and of Teubal (1976) on 'market determinateness' in the Israeli medical electronics industry, both point to the same conclusion. It has been further explored by the interesting work of Mansfield (1977) and his colleagues on the integration between marketing and R & D in project selection systems and the ways in which this influences probability of success.

This interpretation was confirmed by the strong evidence on the occurrence of 'unexpected adjustments' and bugs *after* development in the failure cases, and of the need for user adaptations in nearly half the failures. About three-quarters of the cases of failure showed greater after-sales problems. Thus, 'user-needs understood' is just as much a discriminating measure of efficiency in R & D performance as of marketing and overall management.

Size of *project team* emerged as a clear-cut difference, while other size measures did not differentiate. Since in a number of cases the smaller firms deployed a larger team, this implies a greater concentration on the specific project. This consideration is important in considering the relative advantages of the small firm in innovation in chapter 6. Another measure which strongly suggests the advantages of specialization in R & D is that related to 'coupling' with the outside scientific community.

Carter and Williams (1959) have emphasized good communications with the outside world as one of the most important characteristics of the technically progressive firm. The most backward firms would not of course be found among those attempting to innovate. But among those who were making such attempts there were significant differences in their general pattern of communications. Better external communications were associated with success, but the strongest difference emerged with respect to communication with that *specialized* part of the outside scientific community which had knowledge of the work closely related to the innovation. *General* contact with the outside scientific world did not discriminate between success and failure.

All of these differences may of course be related to the quality and type of management, so that measures relating to the business innovator are perhaps the most interesting. First of all, it should be noted that the business innovator was hardly ever the same man as the chief executive in the chemical industry, but was frequently so in the instrument industry. The most interesting difference between successful and unsuccessful business innovators, and one which was unexpected, was that greater seniority was associated with success. The successful man (they were all men) had greater power, higher formal status, and more responsibility than the unsuccessful. He was also older and had more diverse experience. Some of these differences were not so clear-cut in the instrument industry which may reflect the greater mobility and smaller size of firm, together with more hierarchical structure of management in the chemical industry. Usually the successful chemical innovator had been longer with the innovating firm and in the industry, whereas this was not true in the instrument pairs.

The higher status and greater power of the more successful innovators may be associated with their readiness to take greater risks and to recruit larger teams for their projects. In the chemical industry there was a strong association between success and a more radical technical solution. But taking a variety of measures relating to risk acceptance, there was only very slight evidence that successful innovators assumed greater risks. In the chemical innovations the successful cases were usually the first to market but in the instrument pairs those who came later were usually more successful.

The fact that the measures which discriminated between success and failure included some which reflected mainly on the competence of R & D, others which reflected mainly on efficient marketing, and some which measured characteristics of the business innovator with good communications, confirms that view of industrial innovation as essentially a coupling process, which was suggested at the outset. One-sided emphasis on either R & D or sales does violence to the real complexity of the process. This was strongly confirmed by the multivariate statistical analysis illustrated in Figure 5.1. Composite 'index' variables were formed consisting of several measures relating to one factor. The percentage of points correctly classified was greatest by the combination of the following composite measures: R & D strength, marketing and user needs. In the case of chemicals, composite variables relating to management techniques and management strength were also important. 'Management strength' relates mainly to the status and responsibilities of the business innovator.

The critical role of the 'entrepreneur' (whatever individual or combination of individuals fulfil this role) is to 'match' the technology with the market, i.e.

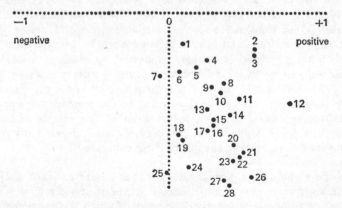

Fig. 5.1 Values of index variable for success points. *Source*: Science Policy Research Unit
(1972)

to understand the user requirements better than competitive attempts, and to ensure that adequate resources are available for development and launch. This interpretation of the key role of the quality of entrepreneurship is in line with the findings of Barna (1962) and of Penrose (1959) and the earlier work of Schumpeter on the theory of the firm (1912, 1942).

In the large firm the business innovator must be high enough in the hierarchy to command resources and get things done. He must have enough knowledge of the way the firm works to know *how* to get things done. In the small firm it frequently means that it will be the chief executive himself, or a man sufficiently close to the chief executive to ensure the necessary concentration of effort. In either case he must be sufficiently powerful and clear about marketing objectives to ensure that the various screening and testing procedures during the course of development and trial production prevent an unsatisfactory product or process coming onto the market. Premature launch may be more dangerous than slowness.

These conclusions will not necessarily be valid for consumer goods innovations where some different mechanisms are at work. In capital goods it is essential to satisfy certain minimal technical performance criteria. The extent to which these generalizations may apply to other industries is discussed in chapter 9. Here it is necessary to consider some other limitations of the analysis, before going on to consider in greater depth the question of size of firm in chapter 6 and the problem of uncertainty, risk and planning in chapter 7.

Thus, the results of SAPPHO confirm points 1, 3, 8, 9 and 10 among the tentative generalizations advanced on p. 112. Point 2 is discussed more fully in chapter 8. Point 4 requires great care in interpretation and the whole of chapter 6 is devoted to this. Points 5 and 7 were not supported by the evidence of SAPPHO. The approach of the innovator to risk again requires much more detailed consideration, and this is attempted in chapter 7, followed by discussion of the implications for theory of the firm in chapter 8.

The generalizations so far made about innovation based on the historical descriptive material in Part One thus find some confirmation from this test. Although they may provide a fairly plausible interpretation of some aspects of industrial innovation since the rise of professionalized R & D, it is certainly not claimed that they are securely based statistically or empirically. The sample was not random and the 'universe' is not known. However, a further sample of fourteen 'pairs' confirmed the original results in all essentials. Moreover, the interpretation of innovation as a 'coupling process' is strongly supported by much additional empirical evidence as well as by logic and common sense. Earlier studies by Carter and Williams (1957, 1959a, 1959b) had led them to formulate the concept of the 'technically progressive firm' embodying many of the combined characteristics of the SAPPHO success cases. Another major series of case studies of industrial innovation in Britain was conducted at Manchester quite independently of SAPPHO at about the same time, and the authors of these concluded:

Perhaps the highest level generalization that it is safe to make about technological innovation is that it must involve synthesis of some kind of need with some kind of technical possibility. The ways in which this synthesis is effected and exploited take widely differing forms and depend not only on systematic planning and the 'state of the art' but also on individual motivations, organizational pressures and outside influence of political, social and economic kinds. Because the innovation process extends over time, it is important to retain continuous sensitivity to changes in these factors and the flexibility to perceive and respond to new opportunities (Langrish *et al.*, 1972).

Additional important empirical evidence for some of the main SAPPHO conclusions came from a Canadian survey of forty-seven new small firms, started by technologically-oriented entrepreneurs. Like SAPPHO this included the study of failures as well as successes. Litvak and Maule (1972) concluded from their survey that:

The marketing performance of the entrepreneurs was weak, and was a major factor for the apparent high mortality rate of the projects. Most of the entrepreneurs were unable to see the linkage between product innovation and marketing innovation. . . . Most of the new product development was carried out and implemented before any attempt was made to assess the market potential and the costs of penetrating the market. . . . The point to be made is that the love that the entrepreneur has for his product innovation often blinds him from perceiving his real opportunities and the state of market competition.

The point about underestimation of user needs and understanding of the market must be heavily underlined, since the SAPPHO inquiry constantly found in discussion with R & D managers and entrepreneurs that they tend to dismiss the point as 'obvious', but nevertheless continue to ignore it in practice.

Further confirmation of some of the SAPPHO conclusions came somewhat unexpectedly from the Hungarian electronics industry (Szakasits, 1974) and from the OECD's international studies of industrial innovation, particularly Pavitt's (1971) cross-country comparisons of the relative innovation success of firms in various member countries, and from the studies of innovation sponsored by the National Science Foundation (NSF) in the United States.

However, even when the statistics are relatively good, as in relation to size of firm (chapter 6), generalizations still need to be heavily qualified. One reason for this is that the 'universe' of innovations or inventions is not known and therefore no strictly random sample can be drawn.[3] Consequently, although attempts may be made to study a representative group of inventions and innovations, as in the Jewkes study or in the research project described in this chapter, we cannot be sure that such a sample is truly representative. This reservation is particularly important when we come to consider so-called secondary or improvement inventions and innovations. There is a tendency in case-study and historical work to concentrate on the more spectacular inventions and innovations. But it can be argued as for example by Gilfillan and Hollander, that the myriad of minor improvements and 'new models' are as important for technical progress as the more radical 'breakthrough' innovation. Moreover, it can also be plausibly maintained that nonspecialists and non-professionals may make a much bigger contribution to the 'secondary type' of innovation than to the breakthrough. It is also probable that knowledge of the market plays a bigger part in this secondary type of invention and innovation than contact with scientific research or, in the case of process innovations, direct experience of operating the process.

Schmookler (1966) showed that in several American industries, over a long period of more than a century, *invention* (as measured by statistics of relevant patent numbers) tended to follow behind *demand* (measured by statistics of investment), with a time-lag of a few years. However, a recent project on the chemical industry at the Science Policy Research Unit showed that in some instances in the

[3] The second Manchester study of innovation (Gibbons and Johnston, 1972) is based on an ingenious attempt to develop a random sample (see p. 159).

chemical industry, there was evidence of 'counter-Schmookler' patterns of growth in the early stages of the emergence of radical new technologies, such as synthetic materials and drugs in the 1920s and 1930s (Walsh *et al.*, 1979). In these periods, surges of inventive activity and of scientific discovery tended to *precede* the take-off of sales and investment, as in the analogous case of the electronic computer discussed in chapter 4. This apparent contradiction may be explained in part if some distinction is made between the more radical inventions and innovations, which are relatively few in number, and the very large numbers of secondary and improvement inventions and innovations, multiplying rapidly as an industry grows and responding directly to market signals and investment behaviour. Thus Schumpeter's theory which puts the emphasis on autonomous innovative activity by entrepreneurs as the mainspring of economic development rather than market demand, can be reconciled with the Schmookler statistics, which measure something rather different.

In so far as patent statistics capture both 'minor' and 'major' inventions, then Schmookler (1966) has shown that professionalized corporate R & D in the 1950s accounted for about half of industrial inventions in the United States and probably for a higher proportion of those which were exploited, i.e. translated into innovations. Many of the other inventions also originated from 'professional' R & D in government and universities. The development and exploitation of inventions emanating from university, government or private inventors probably also involved some professional R & D work in industry in the great majority of cases. But this would still leave a significant number of inventions and innovations which could not be attributed to specialized professional R & D. It is also likely that an even higher proportion of non-patented technical advances are attributable to those outside the professional R & D system.

However, the extent of our ignorance should not be exaggerated. There is firm empirical evidence that most professional industrial R & D is concentrated on product and process improvement and on new 'generations' of established products. What is not known is the relative contribution to technical progress of this R & D work by comparison with the inventions and improvements generated entirely outside the formal R & D system. It is a plausible hypothesis that the proportionate contribution of the formal R & D system is much higher in the research-intensive industries, but it also seems likely that technical progress will be most rapid where there is a very strong interaction between the professional R & D groups and all other personnel associated with the process or product who may themselves contribute to the solution of many problems as well as to their identification. This was confirmed by a detailed study of a major technical innovation in the coalmining industry—the Anderton shearer-loader. Townsend (1976) demonstrated that the highly successful introduction and diffusion of this machine was based on an interplay between a series of more radical inventions and innovations introduced by the machinery-makers (cooperating closely with the Research Establishments of the National Coal Board) and numerous improvement inventions made as a result of operating experience and encouraged by an awards scheme. Both British and German manufacturers contributed to major improvements in the design of this machine, derived in part from their own R & D, and in part from the incorporation of improvements specified by the National Coal Board in Britain. Hollander's work emphasized especially the contribution of the engineering department and Technical Assistance Groups to technical change, but sometimes in association with R & D.[4]

---

[4] This may sometimes be just a question of nomenclature. What is called 'Engineering' or 'OR' or 'Technical department' in one firm may be called 'Process development' or 'R & D in another.

The later stages of 'Project SAPPHO' shifted the emphasis away from the individual cases of success or failure with particular innovations to the study of success or failure of *firms* over a fairly long period. This enabled the project to take account both of individual major innovations and of incremental innovations. Work was concentrated in sectors of the engineering industries, such as textile machinery (Rothwell, 1976), mining machinery (Townsend, 1976) and agricultural machinery (Rothwell, 1979). Whilst firms could and did succeed for short periods by concentrating on incremental improvements, sometimes even without any formal R & D organization, they were often trapped in the long run by an inability to cope with the more radical types of technological competition (such as the Sulzer weaving machine). The results showed that long-run success depended on an ability to combine occasional more radical innovations with a flow of minor improvements in design, responding to customers' wishes and experience. Strong R & D was increasingly necessary to sustain this combination of technical change in the 1960s and 1970s. The earlier SAPPHO case-studies, although oriented towards individual projects, had also pointed in this direction. Especially in the scientific instruments industry, one of the hallmarks of the successful cases was almost always a capacity to incorporate successive design improvements in a series of new models, as for example in the case of the milk analyser (Robertson and Frost, 1978).

One very important piece of empirical work lends support to the view advanced here that specialized R & D and other specialized technical services are increasingly important both for the major radical inventions and innovations and the minor improvement inventions and innovations. This was the work of Katz (1971) in Argentina. He set out to measure the contribution of technical progress to the growth of a large number of enterprises in several branches of the Argentine economy. He was able to collect very comprehensive time series for a large number of firms (250) and to relate his results to measures of the scale of 'adaptive R & D' and other technical activities carried out by the enterprises. From his preliminary interviews he had ascertained that many Argentine firms, whilst not making original radical innovations themselves, nevertheless made many adaptations and improvements to the processes and products which they had acquired either from foreign parent companies or by imitation or licensing. He hypothesized that such 'adaptive R & D' would confer important competitive advantages by enabling the firms to meet the peculiar requirements of the Argentine market more satisfactorily, or to adapt to the specialized operating conditions.

His results showed conclusively that (1) the growth of enterprises was closely related to their technical progress, and (2) their technical progress was strongly associated with the performance of 'adaptive' R & D, and of specialized technical services, although the professional group responsible for this work might be called 'process development' or 'technical department' rather than 'research department'. His results also suggested the important conclusion that 'imitative' or 'adaptive' R & D is more certain in its outcome than 'offensive' or 'defensive' R & D, since studies of firm growth and R & D intensity in the US and UK have not shown such a strong association. Hollander goes so far as to claim that many minor technical improvements are virtually risk-free.

The Federation of British Industries' comparisons of UK firm growth rates and R & D intensity (1947, 1961) did show positive but weak correlations, and fairly strong association at the extremes. These results, taken together with those of the SAPPHO project, suggest that:

1 Firms performing little or no R & D in industries of rapid technical change are likely to stagnate or disappear.
2 Firms performing a great deal of R & D may sometimes enjoy exceptionally high growth rates through offensive success.
3 In the 'defensive' middle zone, variations in R & D intensity show no statistical association with growth, and uncertainty predominates.

Although the statistical association between R & D intensity and subsequent growth by firms is not very strong, the association between *successful innovation* and subsequent growth of the firm *is* strong. Both Mansfield (1968a and b) and other economists have provided convincing confirmatory empirical evidence of the conclusions which common sense suggests—that successful technical innovation leads to the rapid growth of the firm. On the other hand, as we have seen, unsuccessful innovation may lead to bankruptcy, however large the scale of R & D. The implications of the high degree of uncertainty associated with radical innovation are discussed further in the succeeding chapters.

The historical account in Part One suggested that in synthetic materials, in chemical processes, in nuclear reactors and in some electronic systems large firms had predominated in launching the innovations. But a blanket hypothesis of 'bigness wins' could not be sustained, either from Part One or from the SAPPHO project. In scientific instruments in particular, new small firms made outstanding contributions. Inventor–entrepreneurs establishing new firms had apparently been important too in the early days of the chemical industry, and the early days of the semiconductor and radio industries. They have continued to flourish in the minicomputer industry and in computer software. How far is it possible to test generalizations about the relative contribution of large and small firms to industrial innovation?

The evidence from project SAPPHO, so far as it goes, suggests that as between competitive attempts to innovate, size in itself does not affect the outcome very much. However, it is apparent that there is a range of innovations which are not attempted at all by really small firms, so that for example, the competition in the chemical industry or turbine generators is mainly between various large or giant firms. The relative contribution of large and small firms varies a great deal from industry to industry, and investigations such as SAPPHO cannot answer the question of the *aggregate* contribution of large or of small firms to research and innovation in the economy as a whole.

The size structure of industry and its relationship to problems of monopoly and competition is a problem which has preoccupied economists for a long time (Turner and Williamson, 1969) and there is now a considerable amount of statistical information. Unfortunately, in our field of interest most of this relates to R & D, or patents rather than innovation, so that there are big problems of interpretation. This chapter attempts such an interpretation and concludes by reviewing some recent attempts at the direct measurement of the numbers of innovations by large and small firms in British and US manufacturing industry (Freeman, 1971; Kleinman, 1975; and Townsend *et al.*, 1982).

Size of firm and expenditure on R & D

Whereas in the 1950s there was very little reliable empirical evidence outside the US on the degree of concentration in the performance of industrial research and experimental development, such evidence became available in the 1960s. As a result of the efforts of the OECD in standardizing definitions and methods,[1] we now have reasonably comparable data for a dozen countries (OECD, 1969, p. 46). The picture which emerged was consistent and confirmed the hypothesis of those economists who had postulated a high degree of concentration. The hundred largest R & D programmes accounted for more than two-thirds of all industrial R & D in all countries except one, and for more than three-quarters in most cases. The forty largest programmes accounted for more than half of all industrial R & D in all cases except one, and the eight largest for more than 30 per cent in all countries for which figures are available (Table 6.1). In the Netherlands the five largest programmes account for two-thirds of all expenditures (Phillips, Shell, Unilever, AKU, DSM).

---

[1] See Appendix for definitions of R & D, p. 225.

Table 6.1   Percentage of total industrial R & D performed in firms ranked by size of R & D programmes

Country	Number of firms ranked by size						
	4	8	20	40	100	200	300
US	22.0	35.0	57.0	70.0	82.0	89.0	92.0
UK	25.6	34.0	47.2	57.9	69.5	75.0	77.0
France	20.9	30.5	47.7	63.4	81.0	91.2	95.6
Japan	–	–	–	47.7[a]	52.1[b]	63.1[c]	71.4[d]
Italy	46.4	56.3	70.4	81.6	92.5	–	–
Canada[f]	30.3	40.8	58.4	71.5	86.2	93.2	–
Netherlands	64.4[e]	–	–	–	–	–	–
Sweden	33.2	43.0	54.0	71.0	85.4	90.0	–
Belgium	38.5	51.8	72.6	82.7	92.8	97.5	99.4
Norway	29.5	38.8	55.7	70.6	88.2	97.9	100.0
Spain	25.2	47.0	73.9	91.5	–	–	–

[a] The first 54 firms.
[b] The first 85 firms.
[c] The first 180 firms.
[d] The first 289 firms.
[e] The first 5 firms.
[f] Current intramural expenditure.
Source: OECD (1967).

There is less complete but fairly conclusive evidence that the vast majority of small firms in OECD countries do not perform any organized research and development. For France, Britain and the United States, and probably most other countries, the proportion of small firms performing R & D is almost certainly less than 5 per cent (if 'small' is defined as fewer than 200 employees).

It is true that the official statistics of research and experimental development expenditures may not capture research or inventive work which is performed by managers, engineers or other staff incidentally to their main work. It may be that this part-time amateur inventive work is very productive, and the evidence will be discussed for the view that small firms account for an exceptionally high proportion of significant inventions and innovations. But so far as specialized professional R & D activity is concerned, there is pretty firm evidence that this is highly concentrated in large firms in all countries for which statistics are available.

The OECD statistics (Table 6.1) measure the degree of concentration by *size of R & D programme*, and not by size of firm in terms of total employment, turnover or assets. However, for the major countries some statistics are available on concentration by size of firm, although not as a consistent classification. Firms with more than 5,000 employees accounted for 89 per cent of all industrial R & D expenditures in the United States in 1970, and for 90 per cent in 1978. They accounted for about 75 per cent in the German Federal Republic in 1979 and probably about the same proportion in the UK. Firms employing more than 3,000 accounted for about two-thirds of Japanese industrial R & D in 1978–9.

However, the degree of concentration is much less marked by size of firm (classified by total employment) than by size of R & D programme. In the United States there were 466 firms with more than 5,000 employees performing R & D in 1970. But many of them had relatively small R & D programmes, whilst some medium-sized firms (1000–4999 employees) had rather large ones. Thus the 300 largest *programmes* were approximately equivalent to the outlays of the 470 largest *firms*, each accounting for about 90 per cent of the total (Table 6.2). R & D programmes were far more concentrated than sales or employment (Table 6.2). In France the 200 largest *programmes* accounted for about 91 per cent of total expenditures, but the 200 largest firms (measured by employment) accounted for about 72 per cent. There are some industries in which even the largest firms perform little or no research, and others in which even small firms perform a good deal.

Table 6.2   Percentage of R & D, net sales and total employment by companies with largest R & D programmes, US, 1970

Programme size	Total R & D	Federal R & D	Net sales	Total employment
First 4	18	20	6	8
First 8	32	40	9	11
First 20	55	71	16	19
First 40	66	85	23	27
First 100	79	93	38	39
First 200	87	96	50	50
First 300	91	97	63	62

Source: National Science Foundation (1972, pp. 46–7).

The major source of variations in research-*intensity* between firms is the *industry* concerned, so that analysis of the relationship with size is best done industry by industry.

If the vast number of small firms who perform no R & D are *excluded* from the analysis, then any correlation between R & D intensity and size of firm is weak or non-existent. Some economists claim to have discovered inverse correlations at least for some industries in several European countries and for Canada (Hamberg, 1966; Morand, 1970 and de Melto *et al.*, 1980). Both in Japan and the German Federal Republic those few small firms who do perform R & D appear to have an unusually high R & D intensity, but there are some problems over the interpretation of the reporting procedures in the case of the Japanese data.

In addition to the points made at the end of the previous chapter, it could be postulated that the few small firms and many of the medium-sized firms who *do* perform R & D would tend to fall into three categories:

1 Firms which have just begun to develop or exploit a new invention. In this case sales could be relatively low in relation to R & D and a very high research-intensity could be expected. This might tend to fall in the event of successful commercial exploitation of the innovation and growth of the firm and its sales.
2 Highly specialized firms which have a particular expertise, sustained by an intensive research programme in a very narrow field. Here too, research-intensity might often be high.
3 Firms struggling to survive in industries in which new product competition makes R & D increasingly necessary. A very varied management response might be expected in these circumstances, with some firms trying to scrape by with

a sub-threshold R & D effort, others relying mainly on cooperative research, and still others taking high risks with an ambitious programme.

If these suppositions are correct, they would account both for the relatively weak general correlation between research-intensity and size of firm, and for the empirical observations of wide inter-industry variations in the strength of this correlation. In the UK, France, Germany and the US it has been found that in some industries, small or medium-sized firms had higher research-intensity than large firms. Even for all industries taken together, although the United States figures show a consistently higher research-intensity for firms employing more than 25,000, it is the Federal contracts placed with firms that account for the greater part of the difference (Table 6.3). Taking *company-financed* R & D only, the difference by size of firm is not great (remembering that we are dealing here only with those firms that *do* perform R & D). In France, Morand found a *general* inverse correlation by size of firm for those performing R & D, but there are problems with the data because of the inclusion of many small professional associations (1970).

Table 6.3   Funds for R & D as percentage of net sales in R & D performing companies by size of company, US

Firm size	Total R & D (including Federal contracts) as % of net sales			Company funds for R & D (excluding Federal contracts) as % of net sales		
	1957	1967	1977	1957	1967	1977
Less than 1,000	1.8	1.7	1.7	1.4	1.6	1.6
1,000 to 4,999	1.8	1.7	1.5	1.2	1.4	1.3
5,000 to 9,999	} 3.9	2.1	1.9	} 1.6	1.6	1.5
10,000 to 24,999		} 5.2	1.8		} 2.3	1.5
25,000 or more			4.2			2.4
All firms	3.4	4.2	3.1	1.5	2.1	2.0

Source: National Science Foundation.

Turning to variations in research-intensity among large firms, Hamberg (1964) and Scherer (1965a and b) found only a weak correlation with size measured in terms of employment or sales, and still less with size measured in terms of assets. Hamberg's sample consisted of 340 large firms from the *Fortune* '500' list, while Scherer's sample was 448 firms from the same list. Scherer made the interesting observation that in several industries, research intensity generally rose with size up to sales of $250 million, but began to fall somewhere between $200 million and $600 million. (See also the literature review by Kamien and Schwarz, 1975.) Recently, Soete (1979) has examined more recent evidence which has become available during the 1970s and has concluded that the US R & D data do not on the whole support the views of Hamberg and Scherer, although the patent data do still provide some support. Soete maintains that the evidence for the United States supports Schumpeter (Model 2), i.e. there is some tendency for R & D intensity to increase with size of firm with the largest size-groups. This means that the 'hump-backed' distribution noted by Scherer is not characteristic of most industries since the largest firms are generally the most R & D intensive (Table 6.4). Evidence

Table 6.4 Concentration of patents, R & D expenditure, and employment and various inventive activity intensity measures for firms with more than 25,000 employees, ranked by employment

Number of firms included	Percentage of all 130 firms			Number of patents per $ bill. sales	R & D as % of sales	Number of patents per $ mill. R & D
	Patents	Employment	R & D			
First 4	9.04	23.98	24.13	11.86	2.69	0.441
First 8	19.89	34.62	38.39	17.98	2.94	0.609
First 12	25.91	40.84	43.87	20.17	2.90	0.695
First 16	35.21	45.98	51.61	20.06	2.50	0.803
First 20	40.71	50.39	54.50	21.41	2.44	0.879
First 30	53.13	59.28	63.88	24.47	2.50	0.978
First 40	58.31	66.25	69.69	23.03	2.34	0.984
First 50	64.81	71.93	75.11	23.55	2.32	1.015
First 75	78.99	83.87	78.75	23.17	2.14	1.085
First 100	91.08	92.77	94.11	22.99	2.02	1.138
All 130	100.00	100.00	100.00	23.03	1.96	1.176

Source: Soete (1979).

in Part One also suggests that the largest firms were sometimes the most research intensive (IG Farben, Standard Oil and Bell).

Thus, summing up the evidence on size of firm and R & D expenditures:

1 R & D *programmes* are highly concentrated in all countries for which statistics are available.
2 These programmes are mainly performed in large firms with more than 5,000 employees, but the degree of concentration is significantly less by size of firm than by size of programme.
3 The vast majority of small firms (probably over 95 per cent) do not perform any specialized R & D programmes.
4 Among those firms which *do* perform R & D, there is a significant correlation between size of total employment and size of R & D programme in most industries.
5 There is a generally far weaker correlation between the *relative* measure of research activity (research-intensity) and size of firm and it is not significant in some industries and countries.
6 In several countries those few small firms who *do* perform R & D have above average R & D intensities.

Before attempting to interpret these results it is necessary to consider a little further the relationships between R & D expenditures (inputs into R & D) and R & D 'output'.

Size of firm and invention

A number of economists have maintained that despite the heavy concentration of R & D expenditures in large firms, it is the small firms that account for most of the important inventions and innovations. As has already been indicated, whilst the measurement of R & D *inputs* has made significant progress in the past twenty-five years, the same cannot be said of measurement of R & D *outputs*.[2] It was only in 1980 that the OECD devoted a full conference to 'output' (OECD, 1980b). It is generally accepted that the *direct* output of industrial R & D is a flow of information relating to new and improved products and processes. This may take the form of research reports, technical specifications, operational data and instruction manuals based on experience with pilot plants or prototypes, scientific papers, formulae, oral communications, blue-prints, or patents (Table 1.1). No one has found a way to reduce this flow to a common denominator which could be used for inter-firm or inter-industry comparison. The most obvious method would be numbers of 'inventions' and 'innovations', either unweighted or weighted by some kind of qualitative assessment.

The only statistics of numbers of inventions which are generally available are patent statistics, and ingenious attempts have been made to use these for various forms of comparison, including relative output by size of firm. However, as already noted in chapter 3, they are unsatisfactory for a variety of reasons, of which the main one is that firms and industries vary considerably in their 'propensity to patent'. Some firms attach great importance to patents and have large departments with a strong interest in patenting activity, which will tend to 'inflate' their inventive output, when measured in this way. Other firms either do not want to bother with patents or prefer to rely on secrecy. There has been a tendency to assume that large firms would have a higher 'propensity to patent' than small firms and that consequently a measure of output of R & D based on patent statistics would understate the contribution of small firms. Since, in the United States, small firms show a much higher number of patents per dollar of R & D expenditure than large firms, this has been claimed as evidence of superior productivity of small-firm R & D (see Rothwell and Zegveld, 1982).

However, Schmookler (1966, p. 33), the leading expert on United States patent statistics, presented convincing evidence for the view that, contrary to general belief, large firms in the United States have a lower propensity to patent than small ones. He based this on the empirically demonstrable effects of anti-trust actions on the patent policies of large firms, on the far greater possibilities of pretesting before filing of applications of large firms, and on the greater security of large firms in relation to patent-sharing and know-how exchange arrangements. Small firms usually cannot afford *not* to patent and cannot afford to wait, so that patent statistics tend to exaggerate the contribution of the smaller firms to inventive output, and that of private individuals.This view is supported in Britain by the work of Pavitt (1982) and the analysis of the workings of the British patent system by Taylor and Silberston (1973).

The other major problem associated with patent statistics is the variability in importance of patents. One way of trying to get round this difficulty is by weighting patents, or by listing 'major' inventions. The difficulty of these methods is that they are very time-consuming, unless they are confined to a small number of really outstanding inventions. In this case the difficulties which arise are those of subjective judgement in selecting the most important inventions, and of rating the relative importance of radical 'primary' inventions, compared with the vast multitude of secondary 'improvement' inventions. By far the best known example of this technique is Jewkes' study, which has already been discussed (Jewkes *et al.*, 1958)

[2] The whole problem of 'output' measurement in R & D is dealt with more fully in UNESCO (1970), Irvine and Martin (1980, 1981), Martin and Irvine (1982) and Proceedings of OECD Conferences on Output Measurement (1980b and 1982).

and which attempted to show that a majority of seventy major twentieth-century inventions were made *outside* the R & D departments of large firms. The United States Department of Commerce study (1967) adopted an essentially similar view of the importance of private inventors and small firms, but with less empirical supporting evidence, and a tendency to confuse invention with innovation. Similar ideas were propounded earlier by Grosvenor (1929) and more recently by Hamberg (1966).

Jewkes' analysis may be criticized on the grounds that some important corporate inventions were omitted, or perhaps more justifiably, on the grounds that the contribution of large firms has become much more important since the 1920s. If his list of inventions is broken down, the share of corporate R & D is weak before 1930, but dominant since (Freeman, 1967). However, after making allowance for these criticisms, it must be conceded that Jewkes and his colleagues have made a strong case for the view that universities, private inventors and smaller firms have made a disproportionately large contribution to the more radical type of twentieth-century *inventions*. This was also confirmed by our own historical account.

Size of firm and innovation

However, it does not necessarily follow that, because smaller firms may score better on numbers of patents or numbers of 'major' inventions in relation to their R & D inputs, they are consistently more efficient in R & D performance than large firms. First of all, it has already been noted that a number of Jewkes' 'private' inventions were in fact developed and brought to market by large corporations. Of the inventions made outside large firm R & D, perhaps about half were innovated in this way. The final aim of industrial R & D is a flow of *innovations*, so that efficiency in *development* is just as important as the earlier stages of inventive work. Indeed, it is often very difficult to say who made an *invention* because of the tangled chain of claim and counter-claim, but it is usually possible to say more precisely which firms made an *innovation*, in the sense of first launching a new product or process commercially. The relative performance of large firms is apparently better with respect to innovations than with respect to inventions, and Jewkes accepts that their role in *development* work (which is usually far more expensive) is much more important.

Thus, it may be reasonable to postulate that small firms may have some comparative advantage in the earlier stages of inventive work and the less expensive, but more radical innovations, while large firms have an advantage in the later stages and in improvement and scaling up of early breakthroughs. Moreover, there are significant differences between *industries* in the relative performance of small and large firms. In the *chemical* industry, where both research and development work are often very expensive, large firms predominate in both invention and innovation. In the mechanical engineering industry, inexpensive ingenuity can play a greater part and small firms or private inventors make a larger contribution. Patent statistics reflect these differences very clearly, and the point is fully confirmed by the results of the project described at the end of this section. However, it must be noted that in the case of computer software, on which patents cannot usually be taken out, the major contribution of small new firms will not be reflected in patent statistics.

As we have seen in Part One there are some types of innovation which are beyond the resources of the small firm. The absolute number of components is one factor which will affect this. The extreme case is Apollo XI, for which more than two million components were required, but there are other more mundane complex engineering products for which more than 10,000 components may be needed, such as advanced jet aero-engines, electronic telephone exchanges, large computer systems, nuclear reactors, or some process plant. Large firms also have a comparative advantage where there are several possible alternative routes to success, with uncertainty attached to all of them, but benefits from the simultaneous pursuit of several. Similarly, they enjoy an advantage where large numbers of different specialists are needed to solve a problem or expensive instrumentation is essential.

Probably the greatest advantage of the small firm lies in flexibility, concentration and internal communications. SAPPHO suggested that greater concentration of management effort is important. Efficient 'coupling' of marketing-production and R & D decision-making may be much more easily achieved in the small firm environment. In the discussion of the electronic scientific instrument industry reference has already been made to Shimshoni's work (1966, 1970). He found that new small firms had played a critical part in innovating several key instruments and postulated that their main advantages lay in motivation, low costs, lead-time in development work (from speed in decision), and flexibility (Table 6.5). He also concluded that *new* firms had a major advantage in external economies in the form of technological expertise brought from elsewhere in the R & D system. In his studies of 'spin-off' instrument firms, Roberts also pointed to the critical importance of technological entrepreneurs bringing with them ideas and half-developed new products from a scientific environment in university and government laboratories. Golding demonstrated this mechanism operating within the American semiconductor industry. The exceptionally important role played by new small firms and spin-off firms in the American semiconductor industry has led some observers to conclude that Schumpeter's 'Model 1' is a more realistic picture of contemporary reality than his 'Model 2' (i.e. that large firms do not predominate in the process of innovation—see chapter 1). However, before jumping to this conclusion, it is important to keep in mind the following points: first that the larger corporations (Bell, GE, RCA, IBM, etc.) did continue to contribute a large share of the key innovations—perhaps as much as half—throughout the post-war period (Tables 4.4 and 4.6); secondly, that they accounted for more than half the key *process innovations* (Table 4.6b); thirdly, that in Europe and Japan, both the imitation process and the innovation process were dominated to a much greater extent by the large corporations. Rothwell and Zegveld (1982), whilst accepting that small firms enjoy some *advantages* in the innovation process, have also pointed out some of the *disadvantages*, such as access to finance, ability to cope with government regulations and lack of specialist management expertise.

How far is it possible to test systematically the relative contribution of small and large firms to innovation in various industries and the economy generally? Whilst the evidence is still incomplete and the measurement problems remain formidable, there have been several major advances in the 1970s and 1980s and projects carried out in both Britain and the United States enable us to give a fairly definite answer, even if it is not so detailed and precise as we might wish.

A project carried out at the Science Policy Research Unit in 1971 attempted to measure directly the number of innovations made by each of three size-categories of firms in many branches of British industry (Freeman, 1971). The inquiry was carried out for the Bolton Committee of Inquiry on Small Firms, which defined 'small firms' as those with fewer than 200 employees. The survey covered the period 1945–70, but early in the 1980s a second inquiry, supported by the Research

Table 6.5 Comparative advantage of types of firms in instrument innovation

Innovation process	Established large firm	Recent small firm on second or subsequent products	Entrepreneur, first product
Motivation to innovate	3	1—	1
Ability to have or develop own knowledge, technology	1	3	1
Cost advantages, using outside knowledge	2	3	1
Resources available to penetrate market	1	2—	3
Resources for new product development	1	3	1 or 2
Advantage in costs and speed of prototype and early model manufacture	3	1—	1
Flexibility to adopt new product or technology	3	2	1+
Cost advantage, large series production and marketing	1	2—	3

1 = highest comparative advantages, 3 = lowest comparative advantages
Source: Shimshoni (1970, p. 61).

Councils (SERC and SSRC) added new information for the period 1971–80 (Townsend *et al.*, 1982). Lists of important innovations were obtained from independent sources for each of a large number of different branches of industry, comprising altogether about fifty three-digit SIC Minimum List Headings. The industries covered accounted for just over half of total net output of British industry. The innovations were then traced to the innovating firms, 90 per cent of whom were able to supply information on their size in terms of total employment at the time of the innovation.[3]

On the reasonable assumption that the branches of industry included in the survey are representative of British industry as a whole, the most important conclusions were as follows:

1 Small firms accounted for about 12 per cent of all industrial innovations made since the war. This may be compared with their share of production and employment, which in 1963 amounted to about 19 per cent of net output and 22 per cent of employment.
2 The share of small firms in innovation has apparently been fairly steady (Table 6.6), but their share of output and employment has been falling.
3 The share of the largest firms (10,000 employees and over) in the total number of innovations increased substantially over the period, at the expense of medium-sized firms (1,000–9,999 employees).

[3] I am particularly indebted to J. F. Townsend and F. Henwood, who carried out most of the work on this project, and to J. P. Gardiner for additional assistance in the analysis of the relative importance of innovations by small firms.

Table 6.6   Percentage of innovations in each firm size category for each five-year period

No. of employees	1945–9 %	1950–4 %	1955–9 %	1960–4 %	1965–9 %	1970–4 %	1975–80 %	Total %
1–199	16.0	12.0	11.0	11.0	13.0	15.0	17.0	14.0
						(11.0)	(12.0)	(12.0)
200–499	9.0	6.0	8.0	6.0	7.0	9.0	7.0	7.0
						(7.0)	(6.0)	(7.0)
500–999	3.0	2.0	7.0	5.0	5.0	4.0	3.0	4.0
						(4.0)	(3.0)	(4.0)
1,000–9,999	36.0	36.0	25.0	27.0	23.0	17.0	14.0	23.0
						(19.0)	(13.0)	(23.0)
10,000 and over	36.0	44.0	50.0	51.0	52.0	55.1	59.0	52.0
						(59.0)	(66.0)	(54.0)
Total	100.0	100.0	100.0	100.0	100.0	100.0	100.0	100.0
No. of innovations	94	191	274	405	467	401	461	2293

Note: Numbers between brackets for the periods 1970–4 and 1975–80 are the weighted percentage contributions, assuming the same sectoral mix as in the period 1945–69.

Source: J. Townsend *et al.*, *Science and Technology Indicators for the UK: Innovations in Britain since 1945*, Science Policy Research Unit, Occasional Paper No. 16, 1982.

Innovation by size of firm and branch of industry

As expected, the analysis by branch of industry showed big variations in the contribution of small firms to innovation.

Industries may be classified into two fairly clear-cut groups:

1 Those in which small enterprises made little or no discernible contribution to innovation, either absolutely or relatively. These included aerospace, motor vehicles, dyes, pharmaceuticals, cement, glass, steel, aluminium, synthetic resins and shipbuilding (Table 6.7), and (in a special category) coal and gas. In this group small firms accounted for only just over 1 per cent of innovations (6 out of a total of 479), but about 8 per cent of net output in 1963.
2 Those in which small enterprises made a fairly significant contribution to innovation in the industry concerned. These included scientific instruments, electronics, carpets, textiles, textile machinery, paper and board, leather and footwear, timber and furniture, and construction. In this group small enterprises accounted for 103 out of 623 innovations, or about 17 per cent, compared with about 20 per cent of net output in 1963.

If industries are ranked according to the share of small enterprises in the number of innovations for each industry, then this order corresponds fairly well with a measure of concentration based on share of small enterprises in net output (Table 6.7), but the contribution of innovations relative to net output share rises steeply.

In scientific instruments, some types of machinery and paper and board, small enterprises contributed proportionately more than their share of output to innovations. Medium-sized firms (employing 200–999) also contributed substantially to innovation in these industries.

Those industries in which small firms contributed much less than their share of output or nothing at all, correspond broadly to industries of high capital intensity. The major exceptions are aerospace, shipbuilding and pharmaceuticals. In these industries development and innovation costs for most new products are very heavy, although capital intensity is low. Paper and board is again an exception. Although this industry is generally one of relatively high capital intensity, small and medium firms have made important innovations, mainly in speciality products, and in board rather than paper. In these sectors capital intensity is lower.

With this exception it seems to be true that in the capital-intensive industries both process and product innovations have been mainly monopolized by large firms. This finding corresponds closely with the conclusions which emerge from chapters 2, 3 and 4. The small firms make their contribution mainly in the field of machinery and instrument innovations, where both capital intensity and development costs are low for many products, and entry costs are low for new firms. This again corresponds closely to the historical account given in chapter 4. Machinery, instruments and electronics account for two-thirds of all the small firms' innovations reported. Although small firms made a significant contribution to innovation in such traditional industries as textiles, leather and furniture, the total number of innovations in these industries was relatively small. An important conclusion from this inquiry is that Jewkes is right in believing that the growth of professional R & D in the large corporation has not eliminated the contribution of small firms to industrial invention and innovation. But Galbraith is right, too, in believing that the larger corporation predominates in contemporary industrial innovation.

Table 6.7   Share of small firms in innovations and net output of industries surveyed in UK

1958 SIC MLH Number	1958 SIC title of industry	Per cent share of innovation by small firms 1945–70	Number of innovations by small firms 1945–70	Number of innovations by all firms 1945–70	Per cent share of net output by small firms 1963	Value of net output by all firms 1963 (£m)
471–3	Timber and furniture	39	7	18	49	220
351	Scientific instruments	28	23	84	23	154
431–3 450	Leather and footwear	26	5	19	32	157
335	Textile machinery	23	15	65	21	65
481–3	Paper and board	20	6	30	15	317
339	General machinery	17	18	108	14	409
332	Machine tools	11	4	38	18	100
441–15 417, 419 492	Textiles, carpets	10	6	63	18	670
364	Electronics	8	13	160	8	320
211–29	Food	8	3	38	16	814
381	Vehicles, tractors	4	3	64	5	733
276	Synthetic resins and plastics	4	2	52	12	77
370	Ship-building	2	1	59	10	215
271(1)	Dyes	0	0	22	7	35
272(1)	Pharmaceuticals	0	0	44	12	124
463	Glass	0	0	13	14	96
464	Cement	0	0	18	0[a]	41
383	Aircraft	0	0	52	2	185
321	Aluminium	0	0	16	10[a]	100[a]
311–13	Iron and steel	0	0	68	9	630
101	Coal	0	0	23	0	655
601	Gas	0	0	15	0	216
500, 336 337	Construction, earth-moving equipment and contractor's plant	12	4	33	53	1931

[a] Estimated.
Source: Freeman (1971).

As in much industrial economics it is the variation by branch of industry which is important, and the stage of the growth cycle of each industry must also be taken into account. In the early stages of the establishment of a new industry, small innovative firms may be particularly important, but this importance may diminish as the industry matures and economies of scale are exploited. This aspect of the problem is discussed further in chapter 10.

The results of this survey still leave room for argument. The lists of innovations obtained for each industry are obviously still incomplete. In some industries, respondents were able to supply much more complete information than in others. Thus, in ship-building and in iron and steel, more innovations were analysed than in aircraft or aluminium. This was partly because of the exclusion of military innovations, but it was also because of variations in the type of assistance obtained from outside experts. It is very unlikely that a much wider coverage of every industry, and the inclusion of additional branches of manufacturing, would significantly affect the results, but it is just conceivable. It is not claimed that the lists were exhaustive for every branch of industry, nor that some important innovations may not have been omitted. But it is claimed that the lists are representative of major innovations in Britain since the war, and there are no strong reasons to think that small firms are under-represented.

An even more difficult problem relates to the 'weighting' of innovations. It can be maintained that Terylene was a much more important innovation than a traversing creel and that it should be given a correspondingly high weight of, say, 10 or 100. But in practice there is no means of assigning weights to a thousand different innovations. The simplifying assumption is therefore made (as it is by Schmookler in relation to patents) that in dealing with large numbers of innovations these differences tend to cancel each other out.

The small firms' innovations include some very important ones such as plug-board sequence control for machine tools (Nickols Automatics), the printed circuit board for the electronics industry (Technograph) and atmospheric press packing for raw wool fibres (Roypack). All three of these were introduced by new small firms established to exploit the innovation. Among other important small firm innovations are thermo fan drives (Dynair), large diameter bored piles (Economic Foundation and Expanding Piling Company), foam-laminated backings for carpets (Walkden Carpets) and electronic checkweighers (Telemex).

Whilst it was not practical to 'weight' all the innovations in the list, an experimental test was made for several industries. Independent consultants were invited to assign a weight to each innovation for an industry with which they were familiar. The weighting was on a five-point scale and respondents were invited to 'scale' the innovations both for technological and economic importance. The industries chosen were textile machinery, spinning and weaving, and mining machinery; in all of which there were a significant number of small firm innovations. Altogether there were 143 innovations in these industries of which twenty-five were small firm, twenty-nine medium and eighty-nine large firm innovations. The results suggested that a weighting system would tend slightly to diminish the share of small and medium firms in innovation, i.e. by about 1 percentage point for each size group.

Although in general these results confirm the view of Galbraith (1969) that post-war innovation has been dominated by large firms, they also confirm the view of those economists who suggested that the innovative *efficiency* of small firms may be greater than that of large firms (for those few small firms who perform

R & D), in the sense that they apparently produce more innovations per £1 of R & D expenditure than their larger competitors. The share of small firms (as defined in this inquiry) in total R & D expenditures was probably only 3 to 4 per cent, so that they produced a much higher proportion of innovations than their share of R & D, but much lower than their share of total output or employment.

There is support for the view that part of the difference may be attributed to external economies and especially to the mobility of inventor–entrepreneurs and of other scientists and engineers, who bring with them half-developed or fully-developed designs from other R & D laboratories. It is also possible that some of the difference is attributable to the greater proportion of part-time R & D work in small firms, much of which may go unrecorded in formal statistics. Recorded R & D expenditure per full-time R & D employee is much lower in small firms (National Science Foundation, 1972, Tables 7.2, 7.3 and 7.4) and it is likely that this is due partly to underestimation of their expenditures. Finally, it is also possible that large firms suppress or fail to use more of the results of their R & D, thus depressing their 'output' of inventions and innovations.

It is essential to keep in mind the great variety of small firms in industry and to remember that the vast majority do not perform *any* R & D. This innovative small firm is the rare exception, not the general rule. For every small firm engaged in offensive or defensive innovation there are dozens which are 'dependent' satellite firms, and hundreds which follow a 'traditional' strategy—the peasants of industry. They are perfectly capable of efficient production and survival as long as the industries in which they are operating remain relatively unaffected by technical change or alternatively that the results of R & D carried out by others are fairly readily available to all competitors. The building industry is an example of the first, and agriculture of the second. Mechanical engineering, foundries and cotton textiles are examples of industries where it is becoming much more difficult for small firms to survive and there is a high mortality rate among them. These questions of small firm strategy and survival are discussed more fully in chapter 8.

Conclusions

The available international evidence on the whole confirms the findings of the SPRU survey, although with a slightly higher contribution from small- and medium-sized firms than in the UK. For the United States, Kleinman's (1975) results are remarkably similar to those of SPRU (Table 6.8), with the exception of the electronics industry. The differences between the US and the European electronics industries have already been extensively discussed. The contribution of small firms in Japan appears to be even lower than in the UK, despite their higher share of R & D.

Clearly a policy for structure of industry intended to influence the size distribution of firms must be two-pronged. On the one hand, it must both permit and encourage the formation and growth of quite new innovative enterprises launched by inventor-entrepreneurs. The importance of this emerges both from the history of new branches of industry in capitalist countries and from the experience of socialist economies. Commitment to established techniques, institutions and ideas can be a powerful block to change in any society. Dudintsev's novel (1957) and the arrangements made for spin-off firms from the Soviet 'science city' in Siberia both emphasize this and point to the importance of new opportunities for the inventor. The freedom to criticize management and establishment of existing institutions is a necessary condition for innovation, but it can only be really effective

Table 6.8 The size distribution of innovations: UK and US samples compared by percentages of total

Sector		Size of firm by employment			Number of innovations
		1–999	1,000–9,999	10,000 and over	
Chemicals	UK	9.0	7.6	83.3	288
	US	13.1	7.5	79.4	107
Machinery	UK	37.2	41.3	21.5	460
	US	50.0	26.6	23.4	64
Electronic Products	UK	17.4	11.6	71.1	242
	US	40.0	40.9	19.1	115
Instruments	UK	55.4	24.6	20.0	195
	US	56.9	24.2	19.0	211
Electrical Products	UK	25.9	25.9	48.2	81
	US	15.8	12.0	72.3	184
Motor Vehicles and Other Transport	UK	5.3	34.9	59.8	152
	US	7.9	5.3	86.8	38
Total of Samples	UK	25.0	22.8	52.2	2293
	US	30.5	17.7	51.7	862

Note 1: Sectoral breakdowns are according to the innovating firm's principal activity.
Note 2: The sum of the sectors is less than the total as not all sample was included in this comparison.
Source: Townsend *et al.* (1982); for US data, Kleinman (1975).

if new small innovating organizations can set up independently. The possibility of new entrants to an industry is the really effective constraint on monopoly.

However, on the other hand, a romantic policy which relied mainly on encouragement of the small inventor–entrepreneur and on 'trust-busting' would be hopelessly naive and ineffective in coping with most of the contemporary problems of industrial innovation. In many key branches of industry the innovative process is dominated by giant corporations, often American-owned, although operating as 'trans-national' or 'multinational' companies. Their strength is derived from the cumulative experience of R & D and STS which establishes a technical and market know-how lead and is reinforced by static and dynamic economies of scale. These assume increasing importance in those industries where technological complexity imposes high development costs for successive generations of equipment. A comparison of the development costs of the Comet with the SST, the Burton process with fluid catalytic cracking, the ENIAC with the IBM 360 series, the Chicago reactor with the Fast Breeder, or 30 MW turbo-generators with 800 MW, shows an increase by two orders of magnitude or more in each case. In all such industries only a few firms can survive in a competitive world market, but as in the case of the micro-computer, radical technical change may completely transform the situation.

It is essential to realize that the higher the development and associated innovation costs, the greater the advantage to the larger scale producer. Or, to put the matter the other way about, 'the higher the ratio of development to production costs the worse off is the producer with the smaller total market' (Plowden Report 1965).

From this point it is clear what an enormous advantage accrues to manufacturers such as Boeing, IBM, Texas Instruments, GE and Westinghouse for products such as passenger aircraft, large computer systems, nuclear reactors and turbo-generators, and how important a large international market is to them. The development, design and test costs are very high for new generations of equipment and they are an absolute threshold, which must be met by any firm which wishes to compete, irrespective of its sales volume. The dynamic economies of scale associated with the learning curve for a higher volume of production further reinforce this advantage. The predominance of the giant corporation in innovation in many branches of industry is quite understandable in the light of these considerations. This factor is also very important in considering the marketing and licensing policies of multinational corporations in developing countries (Sercovich, 1974).

In R & D intensive industries competition mainly takes the form of technical innovation and technical services to customers. Entry is restricted by R & D capacity and by the need to provide marketing and technical service facilities. Each firm which wishes to stay in the business must be capable, if not of making a major innovation itself, at least of imitating those made by its more advanced competitors within a short time. To do this it must have a certain R & D capacity, even if it also makes use of licensing and know-how agreements. This minimum level of 'defensive' research and development may be termed the 'threshold'. It is an *absolute* level of resources, not a *ratio* of sales. Below this 'threshold' level of R & D expenditure it will normally be impossible to develop new products with lead times short enough to survive. The 'threshold' is low for some types of electronic instruments, and many small firms prosper; indeed, because of their flexibility and speed of reaction, they may have some competitive advantage over larger firms. It is also fairly low for 'mini-computers' but for more complex products, such as communication satellites or electronic telephone exchanges, the threshold is very high.

The 'threshold' figure can be turned into an estimate of the annual R & D expenditure required by dividing it by the lead time and making some adjustment for supporting research in materials, etc. This annual level of expenditure would be the minimum needed to maintain a defensive position in the market, taking into account the lead time of competitors. If a firm's market share is low, then this level of expenditure may be a very high ratio of sales; other firms with a larger market share will have a lower ratio and will make more profits (other things being equal). Thus several leading American firms have a lower ratio of private venture R & D to sales than their British or French counterparts, but because their sales are so much higher they have a far higher absolute level of expenditure. If a firm with a small market share copies the R & D/sales ratio of a more successful competitor, this may well mean that its research expenditure falls below the 'threshold' level.

A national strategy which is designed to ensure survival of nationally-owned firms in these types of world market must take into account these factors realistically. This is the context in which the rationalization of the British aircraft computer, heavy electrical, semiconductor, telecommunications and nuclear industries must be considered, as well as the nationalization of Rolls-Royce. Obviously rationalization in itself is not enough but it was probably a *sine qua non* for survival. If the EEC should prove to be a viable institutional framework, it will increasingly be faced with these problems on a Western European scale. The promotion of 'national' or 'European' firms is of course not necessarily to be preferred to giving free rein to the trans-national corporations. A variety of alternative strategies are possible, including socialization and international ownership or participation in the key world companies in communications, information systems, natural resources and transport. There may also be strong reasons associated with work satisfaction for giving preference to smaller firms. These factors have not been considered here but this does not mean that considerations of social policy should not take precedence over the more directly economic considerations (Schumacher 1973). Further consideration is given to alternative strategies open to firms and governments in chapters 8, 9 and 10.

The power of the giant corporation should not be exaggerated. From everyday observation as well as from Part One, and from the results of SAPPHO and other empirical studies, it is clear that many attempted innovations fail. The assertions which are often made about the *proportion* which fail are rather unreliable for several reasons. Such generalizations are usually based on the experience of one firm or a few firms over a particular period. Moreover, they are usually vague about the criterion of failure. Thus the 'conventional wisdom' of R & D management often refers to a 'success' rate of one project in ten, or even one in a hundred. But everything here depends upon the *stage* at which such measurements are made. The higher figures often refer to the preliminary selection or screening process by which the less attractive R & D *projects* or proposals are weeded out before much money has been spent on them, and long before they reach the stage of commercial launch. 'Shelved' research projects or development projects may be regarded as 'failed innovations' in the early stages (Centre for the Study of Industrial Innovation, 1971a) but the attrition rate is much higher in the R & D stage than after commercial launch. The SAPPHO project was concerned with attempts which reached this last stage.

Nevertheless, the failure rate is still high when it comes to this stage. This chapter discusses briefly the reasons for the high failure rate and some of the main difficulties confronting the firm in its project selection procedures. Finally, it concludes that it will be difficult to reduce this failure rate by better management of innovation or project selection and control techniques, except for the 'adaptive' and 'imitative' type of project.

This conclusion might appear to be at variance with the findings of SAPPHO and other projects designed to increase our understanding of innovation management. But it is important to recognize fully the limitations of such findings. Even if they are broadly correct in their interpretation of the characteristic pattern of success and of failure this is very far from providing a recipe or formula which will *ensure* success.

In so far as the technical and commercial success of other innovators may affect the outcome of each attempt, some failure rate is almost inevitable when there are parallel or competitive attempts. Fuller knowledge about the conditions of success may raise the general standard of management in *all* attempts but it will not eliminate the possibility of failure where winners and losers are part of the game.

An analogy may be made with the management of football teams. Managers of teams are generally aware of what is needed to win a match. They have a fairly good idea of the 'pattern of success'. So usually have their opponents. But it is by no means so easy to translate the ideal into reality on the field of play for many reasons, including the behaviour of the competition. What can be recognized *ex post* cannot always be controlled or initiated *ex ante*. Many of the variables involved are in any case not easy to manipulate.

It is true, of course, that there are some market situations where it is quite possible for several innovators to be successful simultaneously or nearly so. The success of one player does not necessarily mean the failure of another; there are some races where all can have prizes and others which are one-horse races. But even in the case of a monopolist or a socialist system of innovation, failures would persist for three reasons: technical uncertainty, market uncertainty and general business uncertainty.

The last category applies to all decisions about the future and it is generally assumed that a suitable discount rate applied to estimated future income and expenditure is the appropriate way to handle it in project evaluation. However, it will have bigger implications for innovations than for other types of investment to the extent that innovation projects have a longer time scale before the potential benefits can be realized. These implications are discussed at the end of this chapter.

The other types of uncertainty are *specific* to the particular innovation project and cannot be discounted, eliminated or assessed as an insurable type of risk. It is true that technical uncertainty can be very much reduced in the experimental development and trial production stages and that is indeed one purpose of these activities. But the outcome of these stages cannot be known before their completion, otherwise the work is not experimental and the activity is not truly innovative. Moreover, even after successful prototype testing, pilot plant work, trial production and test marketing some technical uncertainty still remains in the early stages of the innovation. As we have seen, one of the characteristics of successful innovators is the effort to get rid of bugs in the development stage. But some usually remain even in well-managed innovations, and occasionally they lead to serious setbacks some time *after* commercial launch. Some very expensive and well-known examples are the Comet jet airliner and Du Pont's Corfam (see chapter 3).

Technical uncertainty is not merely a matter of 'work' or 'not work', although this is, of course, decisive for success. Indeed, the problem is very rarely reduced to this simple level. Much more usually it is a question of *degree*—of standards of performance under various operating conditions and at what *cost*. The uncertainty lies in the extent to which the innovation will satisfy a variety of technical criteria without increased cost of development, production or operation. Several processes for producing indigo dyes synthetically were technically successful without being commercially viable (see chapter 2).

The 'risk' attached to technical innovation differs from 'normal' risks which are insurable. Most economists, following Knight (1965), distinguish between *measurable* uncertainty or risk proper and *unmeasurable* uncertainty or true uncertainty (see also Shackle, 1955 and 1961). Technical innovation is usually classified with the second category. By definition, innovations are not a homogeneous class of events, but some categories of innovation are recognizably less uncertain than others (Table 7.1), and less risky. As Knight recognized, the classification of 'risk' and 'uncertainty' is a matter of degree except in the extremes. Life and fire insurances and other repetitive, calculable risks are usually cited as instances of the first type of 'risk' which can be dealt with in a fairly straightforward manner by the theory of statistical probability, but uncertainty enters in even here. The second type of risk will not normally be assumed by insurance companies or indeed by banks. Special forms of financial institution have, therefore, been developed to handle this kind of uncertainty involving specific judgement in each individual instance.

Even the lower levels of uncertainty illustrated in Table 7.1 are such that only a very small proportion of R & D is financed directly by the capital market at all. Internally generated cash flow predominates. Where the risk is not borne by the firm or those fairly familiar with the individual project, usually either some type of cost-plus R & D contract is needed, or outright government ownership and finance of the R & D facility.

Table 7.1   Degree of uncertainty associated with various types of innovation

1 True uncertainty	fundamental research fundamental invention
2 Very high degree of uncertainty	radical product innovations radical process innovations outside firm
3 High degree of uncertainty	major product innovations radical process innovations in own establishment or system
4 Moderate uncertainty	new 'generations' of established products
5 Little uncertainty	licensed innovation imitation of product innovations modification of products and processes early adoption of established process
6 Very little uncertainty	new 'model' product differentiation agency for established product innovation late adoption of established process innovation in own establishment minor technical improvements

Numerous attempts have been made to deal with the uncertainty inherent in innovation by substituting *subjective* probability or credibility estimates for the relatively *objective* data used in estimating life insurance tables and other insurable risks (for example, Allen, 1968, 1972; Beattie and Reader, 1971). These attempts raise complex philosophical issues which will not be discussed here, but the empirical evidence will be examined to see how firms actually approach innovation decision-making and how far they are capable of statistically-based estimation procedures.

It will be argued that the nature of the uncertainty associated with innovation is such that most firms have a powerful incentive most of the time *not* to undertake the more radical type of product innovation and to concentrate their industrial R & D on defensive, imitative innovations, product differentiation and process innovation. This proposition will be argued on theoretical grounds and the supporting empirical evidence will then be discussed. The distinction between in-house *process* innovation and open market *product* innovation is very important here. *Product* innovation involves both technical and market uncertainty. *Process* innovation may involve only technical uncertainty if it is for in-house application, and, as Hollander has pointed out, this can be minimal for minor technical improvements.

Project estimation techniques and their reliability

Let us now consider the problems confronting the decision-makers in deciding whether to embark on an innovation project in the firm. Basically, they will be concerned, whatever particular selection technique they may favour, or even if they operate purely on 'hunch', to make some estimate of three parameters:

1. The probable costs of development, production, launch and use or marketing of the innovation and the approximate timing of these expenditures.

2. The probable future income stream arising from the sale or use of the innovation and its timing.
3. The probability of success, technically and commercially.

Ideally, the decision-maker would like a complete cash-flow diagram of the future expenditures and income associated with the innovation (Fig. 7.1). Some estimate of the development costs and subsequent launch costs is clearly essential to any kind of assessment of likely profitability, but as we know all too well from the publicized experience of aircraft development, these estimates can often be wildly wrong. Some improvements in estimating *techniques* can be made and a great deal of effort has gone into this, both in military and civil projects. But it must never be forgotten that estimates can only be really accurate if uncertainty is reduced, and uncertainty can only be significantly reduced either by further research or by making a project less innovative. Those firms who speak of keeping development cost estimating errors within a band of plus or minus 20 per cent are usually referring to a type of project in which technical uncertainty is minimal, for example, adapting electronic circuit designs to novel applications, but well within the boundaries of established technology, or minor modifications of existing designs (categories 5 and 6 in Table 7.1).

This conclusion is strongly confirmed by the empirical work which has been done on project estimating errors at the Rand Corporation (Marschak, Glennan and Summers, 1967; Marshall and Meckling, 1962, by Mansfield *et al.*, 1971 and 1977, by Allen and Norris, 1970, Norris, 1971, Keck, 1977, 1980 and 1982, and Kay, 1979). Mansfield's work is particularly valuable because it permits comparison of different types of innovation. Although it must be conceded that it is very hard to measure just how 'radical' any innovation is, it is nevertheless certain that 'new chemical entities' constitute a more radical class of innovations than 'alternative dosage forms', and likewise that 'new products' are normally a more radical departure than 'improved products'. The differences in average cost and time over-runs, as well as in the variance of estimates, are very striking (Table 7.2).

Mansfield's work is also extremely important because it confirms that large errors are not confined to the military sector or the aircraft industry. Moreover, his work on the chemical industry shows that estimating errors cannot be attributed to inexperience, as the firms which he investigated had long experience of project estimation and innovation, and were among the leading R & D performers in the US industry. The results do, however, suggest that there is some trade-off between cost and time, as the average over-run in military projects was much greater with respect to cost than time, whilst the opposite was true of civil projects, both in the US and the UK. The work of Allen and Norris also suggests that time over-runs were greater in *research* than in development. This trade-off has been explored in some depth by Peck and Scherer (1962) in their work on the weapons acquisition process.

In addition to the very large errors involved, the tendency to optimistic bias is notable. This optimistic bias is present in other types of investment forecast, but not in such an extreme form. It suggests strongly that the social context of project 'estimation' is a process of political advocacy and clash of interest groups rather than sober assessment of measurable probabilities. This view is confirmed both by historical accounts of individual innovation decision processes and by what little academic research has been done on the subject. Particularly important here was the work of Howard Thomas (1970) on project estimation in two scientific

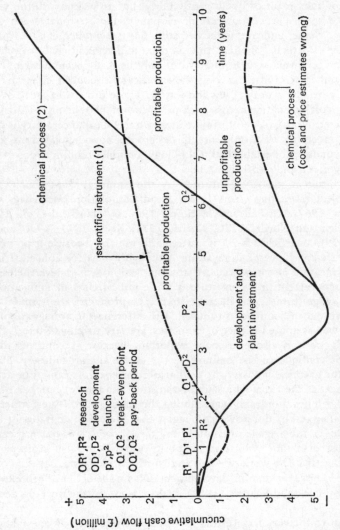

Fig. 7.1   Cumulative cash-flow diagram.

Table 7.2   Average and standard deviation of ratio of actual to estimated cost
by project type and relative size of technical advance, sixty-nine
technically completed projects, US Proprietary Drug Laboratory

Project type	Size of technical advance		
	Small	Large and medium	Total
Product improvement			
Average	1.39	1.49	1.41
Standard deviation	1.39	1.64	1.41
Number of projects	28	5	33
New products			
Average	2.21	5.46	2.75
Standard deviation	3.56	5.86	4.11
Number of projects	30	6	36
Actual to estimated time for above			
Product improvement			
Average	2.80	1.74	2.64
Standard deviation	1.28	0.84	1.27
Number of projects	28	5	33
New products			
Average	3.14	3.70	3.24
Standard deviation	1.53	2.19	1.80
Number of projects	30	6	36

Source: Mansfield *et al.* (1971, pp. 102 and 104).

instrument firms. Not only did he find that engineers deliberately made very conservative estimates of development costs, but he also found that they did this in
spite of strong financial incentives (including profit-sharing arrangements) to make
'honest' estimates:

> Many engineers in the firm admit quite freely that their estimates of cost and
> sales volume for projects are often biased in such a way that the resulting return
> factor estimates appear favourable to the firm. They point out that the pro-
> cedures themselves are very inaccurate and do not incorporate the technical
> feeling about a project that an engineer often has, but is not necessarily under-
> stood by a finance or marketing man. So the engineer deliberately amends
> estimates (the means by which evaluations are made) in order to make the
> return factors acceptable to the firm. They do not do this to make projects
> personally and technically attractive to them more acceptable to the firm,
> because they are aware that the firm's financial interests and theirs are in one-to-
> one correspondence given the profit-sharing and preferential share-purchase
> plans offered by the firm as part of the remuneration package. Their sole motiva-
> tion is to make the firm move towards more flexible numerical criteria for
> differentiating between projects. (Thomas, 1970.)

Whilst it is true that empirical evidence on project estimation is still not as
comprehensive as we might wish, it must be regarded as persuasive support for the
hypothesis that wide margins of error (with an optimistic bias) are characteristic
of the experimental development process. This in itself must make innovation

Table 7.3   Percentage of research managements of US firms rating accuracy
with which factors affecting R & D can be estimated

Factor	1 Excellent	2 Good	3 Fair	4 Poor	5 Totally unreliable	4 + 5
Cost of research	3.5	27.8	52.2	14.8	1.7	16.5
Cost of development	2.6	38.8	46.6	9.5	2.5	12.0
Probability of technical success	3.5	51.3	39.9	6.3	0.0	6.3
Time to complete research	0.9	18.6	50.4	24.8	5.3	30.1
Manpower to complete research	2.6	34.2	53.5	7.0	2.7	9.7
Probability of market success[a]	3.6	33.6	38.2	14.5	10.1	24.6
Time to complete development	1.8	34.5	41.8	17.3	4.6	21.9
Market life of product[a]	4.6	28.0	29.0	23.4	15.0	38.4
Revenue from sales of product[a]	5.3	36.0	28.9	27.2	2.6	29.8
Cost reduction if R & D succeeds	10.7	57.1	14.3	14.3	3.6	17.9

[a] Assuming success of R & D.
Source: Seller (1965), pp. 177–8.

hazardous at least in relation to that part of the decision-making which precedes prototype-test and pilot plant work. The evidence on project estimation in socialist countries also points to the same conclusions (Lakhtin, 1967).

Errors in development cost estimation alone may, of course, be sufficient to bankrupt a firm, if these costs account for a large proportion of its available resources. This occurred in spectacular fashion with Rolls Royce, and several smaller scale examples are cited in Part One. But as we shall see, the market uncertainty is frequently far greater than the technical uncertainty.

Seiler (1965) found that research managements of one hundred large US firms rated 'probability of technical success' and 'development costs' as easier to estimate accurately than either 'probability of market success' or 'revenue from sales of product' (Table 7.3).[1] It is easy to think of several reasons why this should be so:

1. The market launch and growth of sales is more distant in time and may be spread over twenty years. A great many things can change during this time. This is partly a question of general business uncertainty relating to the future, but it is also specific to the project so far as it affects forecasts of consumer behaviour.
2. Whilst the development work is largely or entirely under the firm's own control, this is hardly ever true of the market, particularly in a capitalist economy. Economic theory is not capable of predicting the reactions of oligopolistic competitors in the face of innovation by one member of an oligopoly. Nor can the reactions of future customers or the trends of future legislation in relation to new products be safely predicted.
3. The prediction of future sales revenue and possible profit depends not only on forecasting total quantity which can be sold, but also on forecasting future costs of production, price *and* price elasticity. This is a formidable undertaking for a product not previously used by consumers.
4. Technological obsolescence may kill a new product or process almost as soon as it has been launched.

[1] One can only agree with Mansfield's ironic comments on the apparent optimistic self-deception of US research managements in believing that 'good' estimates could be made of many of these factors. The point here, however, is the *relative* accuracy.

The empirical evidence, although unsystematic, confirms what theory suggests. Early estimates of future markets have been wildly inaccurate. As suggested in Part One, the major civil innovations in the past thirty years have been in the electronic industry and in synthetic materials. Almost every major innovation in these two industries was hopelessly underestimated in its early stages, including polyethylene, PVC and synthetic rubber in the materials field, and the computer, the transistor, the robot and numerical control in electronics.

As has already been shown in detail in chapter 4, one of the most interesting cases is the computer. The early estimates almost all assumed that the market would be confined to a few large-scale scientific and government users. Even firms like IBM, as Watson Junior has confirmed, had no inkling of the potential until several years after electronic computers were in use. 'Optimistic' estimates made in 1955 put the total US computer stock at 4,000 in 1965. The actual figure turned out to be over 20,000. Similarly large errors were made in underestimating the future potential applications of numerically controlled machine tools and robots.

Both of these, together with examples of polyethylene and PVC, are cases of gross underestimation of future market potential for radical innovations. But there are also examples of gross over-optimism, for example in relation to the fuel cell, the airship, 'Ardil' (the synthetic fibre), various nuclear reactors and the IBM 'STRETCH' computer.

It may be said that the forecasting techniques in use before 1960, when many of these estimates were made, were still very primitive and that there are now much more sophisticated techniques which will reduce these errors. This remains to be seen, but the portents are not encouraging. There can have been few cases where more effort and expertise were devoted to market and cost estimation than in the case of Corfam. A computer model of the world market for hides, leather and shoes was developed, and a prolonged programme of manufacturer and customer trials with Corfam uppers. There are few firms with such an impressive record of product innovation as Du Pont, and they probably knew more about the shoe market than any firm in the industry by the time they launched Corfam. Yet they apparently lost about $100 million on this venture before they withdrew the product from the market.

If we consider the various innovations discussed in chapters 2, 3 and 4, it is difficult to think of any which worked out as originally expected. The gestation period was often far longer than the pioneers had anticipated (PVC, ammonia, TV, synthetic rubber, catalytic cracking, optical character recognition, indigo) and the development costs were frequently very much higher. The Concorde was probably the most spectacular example of gross underestimation of R & D costs and overestimation of the market in the 1960s and 1970s, with the result that, despite intense efforts by the British and French governments, they lost over £1,000 million and had to subsidize production, sales and airline operation as well as R & D. Even so, only a very small number of aircraft entered service.

Animal spirits and project estimation

All of this is surprising only to those who believe that some new project evaluation technique or simulation technique would resolve the difficulties which are inherent in the very nature of innovation itself. Keynes was a great deal wiser. Although he was able to make a fortune on the Stock Exchange, and to write a treatise on the theory of probability as well as to revolutionize economic theory, he had no illusions about risky investments, whether speculative or innovative:

> Most, probably, of our decisions to do something positive, the full consequences of which will be drawn out over many days to come, can only be taken as a result of animal spirits—of a spontaneous urge to action rather than inaction, and not as the outcome of a weighted average of quantitative benefits muliplied by quantitative probabilities. Enterprise only pretends to itself to be mainly actuated by the statements in its own prospectus, however candid and sincere. Only a little more than an expedition to the South Pole, is it based on an exact calculation of benefits to come. Thus if the animal spirits are dimmed and the spontaneous optimism falters, leaving us to depend on nothing but a mathematical expectation, enterprise will fade and die;—though fears of loss may have a basis no more reasonable than hopes of profit had before.
>
> It is safe to say that enterprise which depends on hopes stretching into the future benefits the community as a whole. But individual initiative will only be adequate when reasonable calculation is supplemented and supported by animal spirits, so that the thought of ultimate loss which often overtakes pioneers, as experience undoubtedly tells us and them, is put aside as a healthy man puts aside the expectation of death. (Keynes, 1936.)

The uncertainty surrounding innovation means that among alternative investment possibilities innovation projects are unusually dependent on 'animal spirits'. But animal spirits are feared and distrusted by cautious decision-makers. As a standard textbook on investment decisions puts it:

> . . . management will show a preference for projects with known outcomes over those whose outcomes are uncertain and the value of an investment proposal will be reduced according to its degree of uncertainty. (Townsend, 1969.)

Often the substitution of subjectively estimated expected value for an objectively derived probability estimate rests on a false assumption which Keynes exposed, that 'an even chance of heaven or hell is precisely as much to be desired as the certain attainment of a state of mediocrity.' The risk (or uncertainty) aversion of entrepreneurs varies enormously of course, but we are on fairly safe ground in assuming that not many are prepared to gamble their survival on a fifty-fifty chance. Use of subjective probability estimates is usually acceptable only where possible outcomes are not extreme, and there is some repetition of previous experience.

This means that the acceptance of a high degree of uncertainty in innovation is likely to be confined to the following categories:

1. A few small-firm innovators who are ready to make a big gamble, or who are impelled to do so by some threat to their existence.
2. Large-firm innovators who use careful project selection methods but who can afford to adopt a 'portfolio' approach to their R & D, offsetting a few very uncertain investments against a large number of 'mediocre' projects. The size of the very uncertain investments will not usually be such that failure would threaten the continued existence of the firm.
3. Large-firm innovators who are not closely controlled by any formal project selection system and who are able to use corporate resources with a good deal

of freedom, and hence impose their subjective estimates or preferences upon the organization.

4. Large- and small-firm innovators who unwittingly accept a very high degree of uncertainty, through 'animal spirits', because the enthusiasm of inventors, entrepreneurs, or 'product champions' leads them on. In some cases (probably the majority) they may not bother to make any sophisticated calculations of the probable return on the investment. In others they may accept grossly over-optimistic subjective estimates of the probable outcome.

5. Government-sponsored innovators who accept very high risks because of urgent national needs (usually war, or threat of war) or a deliberate national science policy strategy, which creates an assured and profitable market in the event of success.

6. Government-sponsored innovators who accept grossly over-optimistic estimates of future returns for other reasons, where failure does not pose a serious threat to the decision-makers, as in the case of Concorde, where diplomatic and prestige considerations were allowed for nearly twenty years to overrule commercial judgement and commonsense.

The empirical evidence relating to the use of project selection techniques confirms that the more advanced portfolio methods which have been developed by statisticians and management consultants are seldom used. In the US, Baker and Pound (1964) found that a few of the techniques had been used occasionally and then discarded in favour of simpler 'rule of thumb' methods or discounted cash flow (DCF) calculations (see also Rubenstein, 1966). These methods are strongly biased towards short-term pay-back and the system in which they are used is frequently project-based rather than portfolio-based. Their widespread use probably discourages the more radical type of innovation, which would find more favour either in a fairly sophisticated selection system or without any very formal system. A survey in Sweden in 1971 confirmed that only simple quantitative methods were then used in Swedish industry and indicated some reasons for resistance to sophisticated techniques (Naslund and Sellstedt, 1972). Similarly, Olin's survey (1972) concluded that in the European chemical industry project selection remains a pragmatic and intuitive art.

A partial alternative to a quantitative cost-benefit or DCF approach is to use a qualitative check-list method of evaluation. A check-list approach has the advantage of being able to take into account many factors which may be difficult to incorporate in a mathematical formula. For example, a critical factor in the success of any R & D project is the enthusiasm and capacity of the project leader and his other commitments. Another is the firm's resources of skilled people and accumulated know-how in the field and the possible spin-off from other R & D projects. A third may be the firm's relationship with potential customers and so forth. Whilst all these factors may be taken into account by a research manager or an entrepreneur in calculating probability factors for technical or commercial success, a check-list procedure has the merit of compelling fairly systematic attention to be paid to each point. The actual check-list may be varied in accordance with the peculiar circumstances and characteristics of the firm (or other innovating organization). But the kind of questions that would tend to appear in most check-lists would include the following (Dean, 1968 and Seiler, 1965):

1. Compatibility with company objectives;
2. Compatibility with other long-term plans;

   3. Availability of scientific skills in R & D;
   4. Critical technical problems likely to arise;
   5. Balance of R & D programme;
   6. Interaction with other R & D projects;
   7. Competitors' R & D programmes;
   8. Size of potential market;
   9. Factors affecting expansion of the market;
  10. Influence of government regulations and control;
  11. Export potential;
  12. Probable reaction of competitors;
  13. Possibility of licensing and know-how agreements;
  14. Possibility of R & D cooperation with consultants or other organization;
  15. Effect on sales of other products;
  16. Availability and price of materials needed;
  17. Possibilities of 'spin-off' exploitation of innovation;
  18. Availability of production skills and equipment;
  19. Availability of marketing skills and experience;
  20. Advertising requirements;
  21. Technical sales and service provision;
  22. Effects on company 'image';
  23. Risks to health or life;
  24. Probable development, production amd marketing costs;
  25. Possibility of patent protection;
  26. Scale and timing of necessary investment;
  27. Location of new or extended plant(s);
  28. Attitude of key R & D personnel;
  29. Attitude of principal executives;
  30. Attitude of production and marketing departments;
  31. Attitude of trade unions;
  32. Overall effect on company growth.

The more far-sighted firms may also take into account additional external costs and benefits, such as problems of waste disposal or employment effects, re-training requirements and contributions to research outside the company. They may also find even more valuable those types of technique which attempt to foresee and avoid all the conceivable bugs and blockages which could frustrate the future progress of the project (Davies, 1972).

   Although the check-list approach permits consideration of many factors which may be disregarded or overlooked in a quantitative analysis, it too has serious limitations. It does not permit easy comparison between alternative projects or 'ranking' of a list of projects, nor does it provide any indication of the likely absolute size of the pay-off. Since most firms have a back-log of projects from which to choose, these are serious defects. Consequently, the ideal method of project selection is probably a combination of a quantitative cost–benefit approach with a qualitative check-list approach. Several such 'scoring systems' have been developed, and Fig. 7.2 illustrates that originally worked out for project evaluation at Morganite by Hart (1966). Hart's system is based on calculating a project index value, which takes account of estimated peak sales value, net profit on sales, probability of R & D technical success and a time discount factor in relation to future R & D costs according to the formula:

$$I\text{(index value)} = \frac{S \times P \times p \times t}{100\,C}$$

where S = peak sales value £ per annum,
  P = net profit on sales (per cent),
  p = probability of R & D success on a scale 0 to 1,
  t = a time discount factor,
  C = future cost of R & D (£).

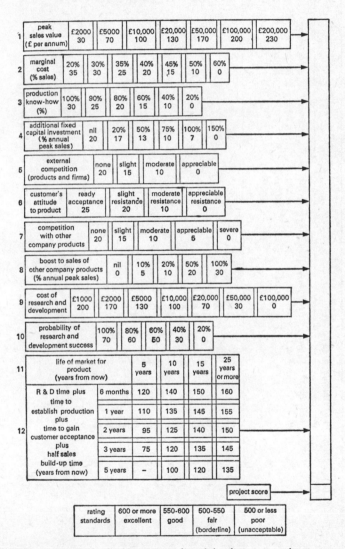

Fig. 7.2 Evaluation chart for product research and development projects.
  Source: Hart (1966).

Estimates for the variables are obtained by answering a check-list of questions. Figure 7.2 shows a score for each possible answer to the question and permits

Table 7.4   Distribution of the working time devoted by researchers to the various activities, analysis of R & D, type of researcher, age, broken down by industry, type of researcher and level of responsibility (provisional)

	Researchers No.	% of time devoted to							
		Individual R & D	Group R & D	Studying and monitoring	Teaching	Organization, administration	Technical services	Advertising, other functions	Other
Food and textiles	187	18	21	11	2	25	16	6	1
Metallurgy	264	11	29	15	2	11	26	7	0
Mechanical Eng.	1,669	23	20	10	2	17	20	6	2
Electrical Eng.	644	17	24	12	2	22	15	6	1
Electronics telecom.	3,056	20	27	12	4	15	16	6	1
Transport vehicles	1,209	22	23	11	3	11	19	8	2
Pharmaceuticals	2,139	22	28	19	3	13	10	3	0
Chemicals	2,328	17	29	14	2	15	14	7	1
Rubber and fibres	868	18	32	12	1	16	15	6	0
Research companies	1,083	22	33	15	3	10	10	5	0
Other manuf. sectors	1,564	16	22	14	2	8	25	9	4
Total	15,010	20	26	13	3	14	16	6	1
% researchers who perform the activity	–	72	83	81	18	60	55	42	6
Scientists	3,248	33	39	14	1	6	4	3	0
Engineers	11,762	16	23	13	3	16	20	7	2
Junior Researcher	5,959	27	26	14	2	8	15	5	3
Project leader	7,154	16	29	13	3	15	17	7	0
Director	1,808	10	18	13	4	27	18	9	1

	Researchers No.	Distribution of R & D time (%)						Type of researcher		
		Funda-mental research	% of re-searchers who per-form the activity	Applied research	% of re-searchers who per-form the activity	Experi-mental develop-ment	% of re-searchers who per-form the activity	Scientists (%)	Engineers (%)	Average age
Food and textiles	187	9	53	34	94	57	94	0	2.4	40.4
Metallurgy	264	2	18	49	94	49	94	1.4	2.0	42.2
Mechanical Eng.	1,669	10	41	56	100	35	83	6.9	6.0	37.0
Electrical Eng.	644	8	28	55	88	37	67	4.1	6.7	31.8
Electronics telecom.	3,056	8	30	41	86	51	85	10.1	20.8	36.2
Transport vehicles	1,209	8	44	51	91	41	82	12.4	11.3	35.9
Pharmaceuticals	2,139	17	62	55	94	28	68	31.2	15.9	36.2
Chemicals	2,328	11	43	43	86	45	83	11.9	14.8	40.5
Rubber and fibres	868	10	39	51	93	39	80	6.0	3.9	38.3
Research companies	1,083	10	47	53	93	38	77	13.8	10.1	35.1
Other manuf. sectors	1,564	8	35	43	98	49	96	2.3	6.0	38.4
Total	15,010	11	43	49	91	41	79	100.0	100.0	37.4
% researchers who perform the activity	–	–	–	–	–	–	–	–	–	–
Scientists	3,248	16	57	69	98	15	58	100.0	–	35.4
Engineers	11,762	9	39	43	89	48	86	–	100.0	37.8
Junior Researcher	5,959	12	46	49	88	38	76	61.5	33.7	34.1
Project leader	7,154	9	40	48	93	43	82	34.2	51.4	38.1
Director	1,808	11	45	50	93	39	83	4.3	14.9	43.4

Source: G. Sirilli, 'The Researcher in Italian Industry', mimeo, 1982.

scoring by addition rather than multiplication by using logarithmic functions of the answers. The method can be varied by using different questions and different scores to suit the circumstances of a particular firm.

The advantage of this method is that it permits the firm to take into account such factors as external competition and customer attitudes, yet at the same time ranks projects on some systematic basis, on the assumption of sales growth and high profitability as major company objectives. Another advantage of the technique is that it can be used to involve all departments of the firm in discussion and evaluation, thereby contributing to mobilization and integration of the firm's resources. This is probably the main benefit of any formal evaluation technique. Indeed, some type of formal technique is usually necessary simply to monitor and control the progress of a project. The periodic revision of estimates and reconsideration of projects is essential for effective management of technical innovations.

However, it is apparent that one of the major factors affecting the selection of a project is the balance of work in the R & D department and in the firm as a whole. Consequently, project selection must be related to programming. What management is looking for is a portfolio of projects rather than a series of separate projects. By thinking in terms of a portfolio rather than a project it is possible to select a blend of 'safe' and 'high risk' projects, so that the more long-term and radical advances are not ignored as they would tend to be if selection were based entirely on a scoring system or rate of return system.

The empirical evidence confirms that industrial R & D is heavily concentrated on the less uncertain types of project. Only a few firms perform any basic research and this accounts for less than 5 per cent of all industrial R & D expenditures in most OECD countries. Several surveys have confirmed that the bulk of R & D expenditures are devoted to minor improvements and quick pay-back projects, rather than the more long-term radical innovations (FBI, 1961; Schott, 1975; Nelson, Peck and Kalachek, 1967). Table 7.4 shows the results of an Italian industrial R & D survey based on data supplied by individual researchers. This confirms the involvement in technical service activity and the relatively low proportion of time spent on fundamental research. It must be remembered that in terms of expenditure (as opposed to time of researchers) the share of experimental development would be higher and that of research lower.

Although not directly related to the distribution of R & D expenditures, the study by Gibbons and Johnston (1972) provides additional interesting evidence. They attempted to derive a random sample of innovations by listing all new product announcements which appeared in UK technical journals on a selected date in 1971. They found that when they examined the list of 1,317 products, after eliminating 258 duplications, 32 process or service innovations, and 16 of non-industrial origin, the remaining 1,000 new products broke down as shown in Table 7.5.

When they came to examine in greater detail a sample of the last 18 per cent, which were those relevant for their purpose, they found that half of this restricted sample could be described as modifications of existing products of the company.

There are grounds for believing that firms are more ready to attempt radical innovations in relation to their own processes than in relation to their products. This includes, of course, the adoption and modification of the *product* innovations of the capital goods industries. The *market* uncertainty is very much reduced with in-house process innovations as the firm controls the application. For similar

Table 7.5

	Percentage
Expansion of range without technical modification	17
New application of existing product	2
Standard product with new specification appropriate to particular application (e.g. portable)	29
Conforming to new standard (e.g. metric)	2
Standard product made easier to use	6
Standard product, new marketing	2
Standard product, new design	2
Products developed outside UK	23
New products, involving technical change and developed by UK firms	18

reasons, much more radical *product* innovations may be expected in response to an assured market (whether government or otherwise) than on a competitive market. This was clearly evident in the development of radar and synthetic materials as well as in military aircraft.

If the process which is developed cannot be used by the firm itself, then the uncertainty is, of course, greater. This difference accounts largely for the respective approach to process innovation of the chemical plant *contractor*, as compared with the chemical firms themselves. As we have seen in chapter 3, the contractors, who cannot use the process themselves but have to face an extremely uncertain market, tend to concentrate on improvements in design and scaling up. They usually attempt completely new processes only in association with a chemical or oil firm. The chemical and oil firms, on the other hand, have a strong incentive not only to make improvements in their in-house processes, but to explore radically new processes for use in their own establishments, hoping that they can keep the innovation sequence largely under their own control. The introduction of such a new process may be used to lower production costs and increase profitability by comparison with competitors, or to lower product prices and expand markets. In some cases introduction may be retarded to preserve the profitability of existing investment. If the process can ultimately be licensed to other firms, this may be regarded as an additional bonus, but the successful marketing (licensing) of the process is not usually essential, as it is in the case of the contractor-originated process. Indeed, there may be a deliberate preference for secrecy and *not* licensing.

R & D budgeting and the strategy of the firm

By undertaking R & D work mainly with a relatively low degree of uncertainty (Table 7.1), the firm is in effect using its R & D budget as a form of insurance against the risks of technical change. Or, as Arrow (1962) puts it, 'the Corporation acts as its own insurance company.' Management often actually bases its R & D budget on a 'percentage of sales' calculation. This 'insurance premium' varies in different branches of industry depending upon the intensity of technological competition, but the level of expenditure is often fairly uniform among many firms in each branch. Although management cannot calculate accurately the return on any individual project or piece of R & D, it has learnt from experience and from observation of competitors that this 'normal' level of R & D spending will probably help it to survive and grow. However, there is room for a variety of alternative strategies and these are discussed more fully in the next chapter. Some firms may spend much more heavily on R & D than is usual for their industry branch and follow a high-risk 'offensive' strategy. Others may try to get by with very little R & D, or none at all, relying on other sources of competitive advantage. Threshold factors complicate the budgeting problem still further, and the factors discussed in chapter 4 illustrate some of the dangers of R & D budgeting by an industry average. Naslund and Sellstedt (1974) produced evidence to show that in Swedish industry many firms allocated funds to R & D on an *ad hoc* project basis, rather than as a stable regular budget. But Kay (1979) argues persuasively that these differences in behaviour may be largely a function of size of firm. The Swedish firms were mainly small ones but the larger European and American firms typically follow a more long-term budget strategy.

Between *industries* wide variations in research-intensity continue to exist, and it may be postulated that they are attributable on the one hand to historical circumstances (new technological opportunities), and on the other hand to the varying pressures of competition. In an industry in which new processes or new generations of products emerge every ten years or so, a moderately high level of research-intensity would be necessary to avoid obsolescence of the product range or excessive costs (drugs, instruments, machinery, vehicles). Although the individual firm in such an industry might increase profitability for a few years by cutting back on R & D, this would be at the expense of long-term profitability and survival. A very low level of R & D activity, or none at all, would be a viable strategy in those branches of industry where technological obsolescence is not a problem, or where changes in product range are mainly fashion-based. An extraordinarily high level of R & D activity might be necessary for survival in industries such as aircraft and electronics where an artificial stimulus to obsolescence derives from military R & D and procurement. In the extreme case, if it works, it's obsolete. The fact that one-third of the net output of the aircraft industry in the United States and the United Kingdom was actually R & D for long periods can only be explained in these 'Alice in Wonderland' terms. Thus the main determinant of research-intensity is the branch of industry, and the ranking of industries by intensity is similar in all industrialized countries.

The outcome of the individual projects with a high degree of technical and market uncertainty cannot be precisely foreseen, either by the firms or by anyone else. Otherwise they would make fortunes more easily and enjoy a high and relatively stable rate of growth. But within an *industry* branch it is much more likely that *someone* will succeed in making the big advances, even though we cannot predict exactly which firm. Thus the uncertainty attached to R & D would lead one to expect a stronger statistical association between R & D spending and the growth of an *industry* than between R & D spending and the growth of the individual firm. This is in fact what the empirical evidence does show.

The most research-intensive industries are, by and large, those with the highest growth rate (Fig. 1.2), while industries with little R & D are on the whole relatively slow-growing or stagnant. But *within* a fast-growing research-intensive industry, such as electronics or pharmaceuticals, there is not such a strong association between high growth and research-intensity by firm. Some of the empirical data suggests a weak correlation and some suggests none at all. This result could be expected not only on grounds of the high degree of uncertainty surrounding the outcome of expensive 'offensive' projects in any individual firm, but also on grounds of externalities which continually arise in R & D. Many firms in an industry may benefit from the technical progress made in only one or two or in a different industry altogether. The whole electronics industry benefited from Bell's work on semiconductors, but only a small part of this benefit was recovered by Bell in the form of licence and know-how payments or indeed in sales. Whilst the patent system strengthens the possibility of appropriating the benefits of information gained through in-house R & D, it cannot and does not prevent the diffusion of this information through a variety of channels, particularly the movement of people (Fig. 4.1, pp. 97–9).

A much stronger association might be expected between firm growth and some combined measure of R & D with Scientific and Technological Services (STS). This would capture the important productivity growth attributable to minor productivity improvements of the type described by Hollander. The empirical results of Katz confirm this hypothesis and so too does some work on the distribution of qualified manpower and growth of firms.

Firms recognize the need to perform R & D in order to stay in business or retain their independence, but there is no recipe for successful innovation. This is one of the main factors contributing to 'higgledy-piggledy' or 'Tolstoyan' patterns of growth. Among the characteristics of successful innovation are the capacity of the innovator to *couple* efficient R & D with knowledge of the market requirements. But this is more obvious *ex-post* than *ex-ante*. Burns and Stalker in a classic study (1961) have shown the internal difficulties within the firm in achieving the necessary degree of integration of these functions. Since innovation is a complex of events extending over several years, the coupling process is a continuous one and is liable to be severely strained by internal problems within the firm as well as by extraneous events. The process is one of groping, searching and experimenting and even the best-laid plans may come to grief. A firm with an efficient R & D set-up is more likely to survive, but it is by no means sure to do so. Even its own innovations may increase the general instability and uncertainty, so that they will often be unwelcome within the firm itself. Project SAPPHO showed that typically there was opposition to an innovation within the firm both on commercial and technical grounds. A greater degree of opposition on commercial grounds was quite strongly associated with failure of the innovation.

The existence of conflict in relation to R & D decision-making and the uncertainty inherent in the process mean that selection and forecasting procedures are not always what they appear from formal descriptions of the methods. A great deal has been written about various modes of 'technological forecasting' and their application in American industry (Jantsch, 1967). Such TF techniques can undoubtedly be very useful and Bright (1968) in particular has demonstrated their value in company strategy in identifying new technological opportunities and threats. However, as with other management techniques, the reality differs from the impression given by enthusiastic advocates. In view of the importance of recognizing what *really* happens in industry, and distinguishing this from idealized abstract concepts, the conclusions of one survey are quoted at some length:

> Since some companies could probably benefit from formal technological forecasting but do not practice it, we searched within the organization for factors that inhibit its use. We found these common management-oriented obstacles to the use of this technique:
>
> 1. *Failure to integrate technological forecasting into the organization's regular plans.* While most managers support the viewpoint that the most critical factor in implementing any forecasting technique is its integration into a long-range

planning program, including the selection of research projects and the allocation of resources consistent with overall corporate objective, this is most often not the case in practice.

More typical is the experience of the executive who was transferred into the advanced planning group of his company with the task of instituting a formal forecasting program to simplify the planning process. . . . There had been no attempt to apply forecasting to the technological future of the company's major product line, and hence his efforts had had no impact on planning.

In another company, one individual with a technical background developed an interest in sophisticated technological forecasting and, with the support of the corporate vice-president for research, had been developing descriptive reports for more than a year. In addition to preparing reports on techniques, he also addressed the R and D planning process, with special attention to the problems of integrating technological forecasting with planning. Yet we found no evidence that anyone was using these techniques for decision making contrary to reports by Jantsch on the same organization. The company's efforts represent the work of one man who had hopes for the future, but has met with little success to date in selling his ideas.

2. Failure to objectively select research and development projects. In most of the companies we studied, the planning and control of R and D expenditures appears haphazard at best. An objective, factual assessment of the economic benefits, direct and indirect, of R and D investments seems to be very much the exception. While part of this is a natural outgrowth of the inevitable uncertainty of the task and the necessary flexibility and informality that characterize most research activities, it also represents management's failure to deal adequately with the planning and control process. The R and D project-selection process observed was primarily one of 'advocacy', based on the personal interests of researchers, the pet projects of key administrators, and a variety of other criteria which could be at odds with the strategic interests of the company.

In one company major R and D decisions are determined by internal power dynamics, which has led to a considerable amount of 'hobby work', or unauthorized research on pet projects. In another, funds are allocated by function or discipline on the basis of advocacy and power, even though it is recognized that individual product allocations might provide a more solid basis for planning. Despite the need for an objective cost-justification of research projects, we found little incentive among R and D decision makers for either planning or forecasting of technology.

3. Failure to understand the role of sophisticated management techniques. A further aspect of managerial resistance to technological forecasting (and to other management techniques) results from a fear of the unknown, a concern that decision-making prerogatives are being pre-empted, and/or the fear that systematic decision-making techniques may uncover incorrect decisions made in the past. In addition, the adoption of sophisticated forecasting is likely to further complicate the planning task rather than simplify it.

4. Failure of top management to support forecasting efforts. The support of top management is a requisite for many major changes, but we found few top managers supporting technological forecasting, and none initiating it. Initiation generally came from one man with the right background, interest and motivation, but without the influence necessary to establish his technological forecasting ideas.

The staff-line barrier is another aspect of this problem, manifested by a corporate staff trying to sell the technique to a divisional planning group, which, in turn, is asked to sell it to the divisional management.

5. Failure of divisional management to look far enough ahead. A final management impediment to technological forecasting is the short time perspective of line decision makers in profit-controlled divisions. The pay-off from technological forecasting is often in the long run, and, as one director of technology planning noted, 'The big corporation has no memory for the long-term investment' (Dory and Lord, 1970; Roberts, 1968).

Many other empirical surveys could be cited to confirm these conclusions, notably Olin (1972). They support the general arguments of Nelson and Winter (1977), Downie (1958), Gold (1971 and 1979) and Marris (1964) on the theory of the firm. Downie was concerned to explain why it was that the process of concentration in more efficient firms did not proceed more rapidly and more 'rationally' since big inter-firm differences in efficiency were clearly apparent. His explanation was in terms of 'unexpected' success of innovations in firms which had fallen behind. Thus the 'innovation mechanism' offset the efficiency 'transfer mechanism', constantly changing the relative position of competitive firms. Marris was arguing that growth maximization was a more realistic explanation of firm behaviour than profit maximization. He postulated, however, that such growth policies were subject to a profits constraint. In so far as R & D is regarded mainly as a force contributing to growth and survival, its spread may be associated with the type of professional management attitudes which he identifies as characteristic of the modern corporation, but in so far as they are unable to estimate likely profitability, it will remain higgledy-piggledy, with the profit/survival constraint sometimes weeding out the unlucky as well as the inefficient.

Moreover, the empirical evidence confirms that decision-making in relation to R & D projects or general strategy is usually a matter of controversy within the firm. The general uncertainty means that many different views may be held and the situation is typically one of advocacy and political debate in which project estimates are used by interest groups to buttress a particular point of view. Evaluation techniques and technological forecasting, like tribal war-dances, play a very important part in mobilizing, energizing and organizing.

Although a 'defensive' R & D strategy may be regarded as the typical response of the firm in research-intensive branches of industry, it is by no means the only possible response. Particularly in countries where science is in any case underdeveloped, such a strategy may not be a realistic possibility. Even in the case of firms in 'advanced' economies, managements of some firms may prefer a strategy of 'imitation', 'dependence' or even suicide. These may be the only realistic alternatives where stong military or civil demands from government provide a powerful artificial stimulus in particular sectors of the world market, or where multi-national corporations enjoy overwhelming advantages in scale of R & D, dynamic economies and market power.

The possibilities open to the firm will be considerably affected by national innovation policies. From the historical account in Part One it was evident that the growth of synthetic materials in Germany and of electronics in the US were intimately related to government policies. By greatly reducing both the technical and the market uncertainty, governments provided a very powerful stimulus to industrial innovation. The profitability constraint (or profit-maximizing behaviour for those who prefer this assumption) means that the time horizon of most firms in their decision-making is relatively short. This inevitably militates against long-term strategies, so that the advocates of long-term R & D policies in the firm will usually be at a disadvantage, unless they have external support of this kind. This is of great importance in considering such problems as pollution, energy conservation and resource depletion.

For this reason (the uncertainty and the long-term nature of radical R & D), it must be expected that there will be a tendency in a private capitalist economy to under-investment in long-term research and innovation, in spite of the potential advantages which the individual firm may gain. This under-investment will be greatest in fundamental research and the more radical types of innovation (Nelson, 1959; Arrow, 1962). It is largely for this reason that in capitalist and socialist economies alike, governments finance most fundamental research and a certain amount of 'radical innovation'. Conversely, there may well be over-investment in short-term R & D associated with product differentiation and 'brand image'. In conditions of oligopoly the firm strives to reduce market uncertainty by differentiating the market for its own products through a combination of advertising and minor technical changes. It does not necessarily follow, however, as Arrow and other economists have suggested, that a more 'perfect' innovation system would separate R & D from the firm. The evidence from SAPPHO as well as most other innovation studies confirms the view that the 'coupling' entrepreneurial function of the firm can be most efficiently performed if it is active in R & D itself. Successful innovation depends on *combining* technical with market knowledge. There is also the negative evidence that those socialist economies which initially separated industrial R & D from the enterprise were generally revising these policies in the 1970s in the direction of greater emphasis on enterprise-level R & D. However, the *balance* between central government R & D and enterprise level R & D is a difficult problem, and it will be argued in chapter 9 that there is room for an increasing role for government laboratories, provided the 'coupling problem' is constantly kept in mind.

The overall picture which emerges from this survey of uncertainty and project selection in relation to innovation is rather more Tolstoyan than the neo-classical theory of the firm tends to assume. Most firms are unable to make very rational calculations about any one project, because of the uncertainty which is inherent in the process, because they lack the information necessary for rational behaviour and because they lack the time and the inclination to get it or to use very complex methods of assessment. This means that growth is higgledy-piggledy and that no one foresees very clearly the outcome of his own or his competitors' behaviour. If anyone doubts this let them consider the behaviour of the firms involved in the United States and European computer industries between 1950 and 1980, or in the radio industry between 1900 and 1930. Nevertheless, the social benefits and costs arising from this untidy innovative process can be very great. We now turn to a consideration of the various strategies which are open to the firm in the face of this degree of technical and market uncertainty, and in the final chapter we consider the problem from the standpoint of national policy.

Even though the survival and profitability constraints are obviously of the greatest importance in explaining firm behaviour, we conclude from chapter 7 that rational profit-maximizing behaviour (or wealth-maximizing) is seldom possible in the face of the uncertainties associated with *individual* innovation projects. This is not to deny that neo-classical short-run theory is a valuable, precise, abstract model of firm behaviour, but it means that this model has limited relevance, and that other ways of interpreting and understanding innovative behaviour are needed (Nelson and Winter, 1977 and 1982). One possible approach to such a theory (and it is no more than a first approach) is to look at the various *strategies* open to a firm when confronted with technical change. Such an approach does not look to an equilibrium which is never attained, but does take into account the historical context of any industry in a particular country. This chapter classifies some possible strategies, and discusses them in relation to R & D, and other innovative activities of the firm.

Any classification of strategies by 'types' is necessarily somewhat arbitrary and does violence to the infinite variety of circumstances in the real world. The use of such ideal 'types' may nevertheless be useful for purposes of conceptualization, just as the use of the concepts of 'extrovert' and 'introvert' is useful in psychology. In practice there is an infinite gradation between types, and many individuals possess characteristics of both types. Moreover, individuals (and firms) do not always behave 'true to type'. Finally, people and firm strategies are always changing, so that generalizations which were true of a previous decade will not necessarily be true of the next.

Any firm operates within a spectrum of technological and market possibilities arising from the growth of world science and the world market. These developments are largely independent of the individual firm and would mostly continue even if it ceased to exist. To survive and develop it must take into account these limitations and historical circumstances. To this extent its innovative activity is not free or arbitrary, but historically circumscribed. Its survival and growth depend upon its capacity to adapt to this rapidly changing external environment and to change it. Whereas traditional economic theory largely ignores the complication of world science and technology and looks to the market as *the* environment, changing technology is a critically important aspect of the environment for firms in most industries in most countries.

Within these limits, the firm has a range of options and alternative strategies. It can use its resources and scientific and technical skills in a variety of different combinations. It can give greater or lesser weight to short-term or long-term considerations. It can form alliances of various kinds. It can license innovations made elsewhere. It can attempt market and technological forecasting. It can attempt to develop a variety of new products and processes on its own. It can modify world science and technology to a small extent, but it cannot predict accurately the outcome of its own innovative efforts or those of its competitors, so that the hazards and risks which it faces if it attempts any major change in world technology are very great.

Yet not to innovate is to die. Some firms actually do elect to die.[1] A firm which

[1] Metcalfe's study (1970) on Lancashire cotton firms showed that a large number were not

fails to introduce new products or processes in the chemical, instruments or electronics industries cannot survive, because its competitors will pre-empt the market with product innovations, or manufacture standard products more cheaply with new processes. Consequently, if they wish to survive despite all their uncertainties about innovation, most firms are on an innovative treadmill. They may not wish to be 'offensive' innovators, but they can often scarcely avoid being 'defensive' or 'imitative' innovators. Changes in technology and in the market and the advances of their competitors compel them to try and keep pace in one way or another. There are various alternative strategies which they may follow, depending upon their resources, their history, their management attitudes, and their luck (Table 8.1).

They differ from those which are normally considered in relation to the economist's model of perfect competition, since two of the assumptions of this model are perfect information and equal technology. Both of these assumptions are completely unrealistic in relation to most of the strategies we are considering, but they are perhaps relevant for the 'traditional' strategy which may be followed by firms producing a standard homogeneous commodity under competitive conditions. Such firms can concentrate all their ingenuity on low-cost efficient production and can ignore other scientific and technical activities or treat them as exogenous to the firm. Some products are still produced under conditions which may sometimes approximate to traditional competitive assumptions but they are only at one end of a spectrum. The 'traditional' strategy is essentially non-innovative, or in so far as it is innovative it is restricted to the adoption of process innovations, generated elsewhere but available equally to all firms in the industry. Agriculture, building and catering are examples of industries which in some respects approximate to these assumptions.

We consider six alternative strategies, but they should be considered as a spectrum of possibilities, not as clearly definable pure forms. Although some firms recognizably follow one or other of these strategies, they may change from one strategy to another, and they may follow different strategies in different sectors of their business.

'Offensive' strategy

An 'offensive' innovation strategy is one designed to achieve technical and market leadership by being ahead of competitors in the introduction of new products.[2] Since a great deal of world science and technology is accessible to other firms, such a strategy must either be based on a 'special relationship' with part of the world science–technology system, or on strong independent R & D, or on very much quicker exploitation of new possibilities, or on some combination of these advantages. The 'special relationship' may involve recruitment of key individuals, consultancy arrangements, contract research, good information systems, personal links, or a mixture of these. But in any case the technical and scientific information

willing to purchase a simple new piece of equipment (a size box), even though it cost less than £100, and the pay-back period was clearly demonstrated by the Research Association and the manufacturers to be less than one year. Mansfield's study (1971) of the adoption process of numerically-controlled machine tools in the American tool-and-die industry similarly showed that many firms did not intend to adopt, 'even when firm owners granted that the lack of numerical control would soon be a major competitive disadvantage'. Mansfield estimated the median pay-back period in this case as five years and suggests that in many firms in this category the owners were close to retirement.

[2] The new product may, of course, be a 'process' for other firms.

Table 8.1   Strategies of the firm

Strategy	In-house scientific and technical functions within the firm									
	Fundamental research	Applied research	Experimental development	Design engineering	Production engineering quality control	Technical services	Patents	Scientific and technical information	Education and training	Long-range forecasting and product planning
Offensive	4	5	5	5	4	5	5	4	5	5
Defensive	2	3	5	5	4	3	4	5	4	4
Imitative	1	2	3	4	5	2	2	5	3	3
Dependent	1	1	2	3	5	1	1	3	3	2
Traditional	1	1	1	1	5	1	1	1	1	1
Opportunist	1	1	1	1	1	1	1	5	1	5

Range 1–5 indicates weak (or non-existent) to very strong.

for an innovation will rarely come from a single source or be available in a finished form. Consequently the firm's R & D department has a key role in an offensive strategy. It must itself generate that scientific and technical information which is not available from outside and it must take the proposed innovation to the point at which normal production can be launched. A partial exception to this generalization is the new firm which is formed to exploit an innovation already wholly or largely developed elsewhere, as was the case with many scientific instrument innovations. The new small firm is a special category of 'offensive' innovator. The remarks here apply primarily to already established firms, but we may recall the conclusion of chapters 6 and 7 that the importance of the new small innovating firm is related to the reluctance and inability of many established firms to adopt an offensive strategy.

The firm pursuing an 'offensive' strategy will normally be highly 'research-intensive', since it will usually depend to a considerable extent on in-house R & D. In the extreme case it may do nothing but R & D for some years. It will attach considerable importance to patent protection since it is aiming to be first or nearly first in the world, and hoping for substantial monopoly profits to cover the heavy R & D costs which it incurs and the failures which are inevitable. It must be prepared to take a very long-term view and high risks. Examples of such an offensive strategy which have been considered in Part One are RCA's development of television and colour television, Du Pont's development of nylon and Corfam, IG Farben's development of PVC, ICI's development of Terylene, Bell's development of semiconductors, Houdry's development of catalytic cracking, and the UK Atomic Energy Authority's development of various nuclear reactors. It took more than ten years from the commencement of research before most of these innovations showed any profit, and some never did so.

The extent to which an offensive strategy requires the pursuit of in-house fundamental research is a matter partly of debate and partly of definition. From a narrow economic point of view it is fashionable to deride in-house fundamental research, and to regard it as an expensive toy or a white elephant. Certainly it can be this, and the advice of many economists and management consultants to leave fundamental research to universities has a kernel of good sense. But it may be too narrow. Certainly some of the most successful 'offensive' innovations were partly based on in-house fundamental research. Or at least the firms who were doing it described it as such, and it could legitimately be defined as research without a *specific* practical end in view (the definition of applied research). However, it was certainly not completely pure research in the academic sense of knowledge pursued without *any* regard to the possible applications. Perhaps the best description of it is 'oriented fundamental research' or 'background fundamental research'. A strong case can be made for doing this type of research as part of an offensive strategy (or even in some cases as part of a defensive strategy).

The straightforward economic argument against in-house fundamental research holds that no firm can possibly do more than a small fraction of the fundamental research which is relevant, and that in any case the firm can get access to the results of fundamental research performed elsewhere. This over-simplified 'economy' argument breaks down because of its failure to understand the nature of information processing in research, and the peculiar nature of the interface between science and technology. There is no direct correspondence between changes in science and changes in technology. Their interaction is extremely complex and resembles more a process of mutual 'scanning' of old and new knowledge. The argument that 'anyone can read the published results of fundamental scientific research' is only a half-truth. A number of empirical studies which have been made in the United States indicate that access to the results of fundamental research is partly related to the degree of participation (Price and Bass, 1969). Many case studies of innovation show that direct access to original research results was extremely important, although the mode of access varied considerably (Illinois Institute of Technology Research Institute, 1969; Langrish *et al.*, 1972; Wilkins, 1967; Gibbons and Johnston, 1974; Industrial Research Institue Research Corporation, 1979). In-house fundamental research was obviously important in some of the cases considered in Part One (e.g. nylon and polyethylene), and its role in relation to Bell's discovery and development of the transistor is discussed in a classic paper by Nelson (1962). It was also important in a significant proportion of the American case studies, for example in GE and Dow. The results of SAPPHO, although not strongly differentiating between success and failure on the basis of fundamental research performance, did suggest a marginal advantage to fundamental research performers (Science Policy Research Unit, 1971 and 1972). It may sometimes be a matter of hair-splitting as to whether research is defined as 'background', 'oriented basic' or 'applied research'. The difficulties in defining and measuring the various categories of R & D are discussed more fully in the Appendix, but it must always be remembered that all schemes of classification are to some extent arbitrary and artificial.

Price and Bass (1969) have attempted to measure the relative importance of direct participation as one of the modes of access to original research. They classified 244 'coupling events' in twenty-seven innovation case studies. A 'coupling event' is one which links developments in basic science with technological advances. The results shown in Table 8.2 indicate that 'direct participation' was involved in

Table 8.2   Frequency of use of coupling method

Category of coupling	Suits and Bueche	Frey and Goldman	Tanenbaum (MAB)
Indirect[a]	8	5	25
Passive availability[b]	28	17	43
Direct participation[c]	38	18	40
'Gatekeeper'[d]	14	2	6
All 'coupling events'	88	42	114

[a] No direct dialogue between originators and users of new scientific knowledge.
[b] Scientists are open to approach but do not initiate a dialogue. Technologists request assistance.
[c] Includes inter-disciplinary teams, exchanges and consultants.
[d] Gifted individuals assigned the specific function of promoting communication between scientists and engineers.
Source: Price and Bass (1969).

forty per cent of the 'events', and 'passive availability' of scientists outside firms was also very important. It is not unreasonable to postulate that here too the effectiveness of communication is to some extent a function of the degree of involvement in basic research.

Most of these studies relate to innovations made by firms which would probably be classified as 'offensive', and tend to confirm the view that in-house oriented fundamental research combined with monitoring activities and consultancy are important modes of access to new knowledge for firms pursuing such a strategy. Price and Bass conclude that:

1. Although the discovery of new knowledge is not the typical *starting point* [my italics] for the innovative process, very frequently interaction with new knowledge or with persons actively engaged in scientific research is essential.
2. Innovation typically depends on information for which the requirement cannot be anticipated in definitive terms and therefore cannot be programmed in advance; instead key information is often provided through unrelated research. The process is facilitated by a great deal of freedom and flexibility in communication across organizational, geographical and disciplinary lines.
3. The function of basic research in the innovative process can often be described as meaningful dialogue between the scientific and technological communities. The entrepreneurs for the innovative process usually belong to the latter sector, while the persons intimately familiar with the necessary scientific understanding are often part of the former.

These findings are extremely important, because it has often been concluded from individual case studies that technical innovations bear no relation to basic research or the advance of scientific knowledge. The results of the American Department of Defence 'Project Hindsight' (Sherwin and Isenson, 1966) and of the Manchester 'Queen's Award' study (Langrish *et al.*, 1972) were often wrongly construed in this way, because they suggested that most of the new products were based on an 'old' science. Any major innovation will draw on a stock of knowledge, much of which is 'old' in this sense. But the capacity to innovate successfully depends increasingly on the ability to draw upon this whole corpus of structured knowledge, old and new.

The availability of external economies in the form of a highly developed scientific and technological 'infrastructure' is consequently a critical element in innovative efficiency. Although these external economies are to some extent world wide, and to this extent it makes sense to talk of a world 'stock' or 'pool' of knowledge, access to many parts of it is limited. Cultural, educational, political, national and proprietary commercial barriers prevent everyone from drawing freely on this stock as well as purely geographical factors. The ability to gain access to it is an important aspect of R & D management and bears a definite relationship to research performance and reputation. Pavitt's inter-country comparisons of innovative performance (1971 and 1980) also bear out this conclusion and so too does the study by Gibbons and Johnston on the interaction of science and technology (1974).

We may conclude, therefore, both from the results of Price and Bass and from our own survey, that the performance of fundamental research, whilst not essential to an offensive innovation strategy, is often a valuable means of access to new and old knowledge generated outside the firm, as well as a source of new ideas within the firm. Whilst ultimately all firms may be able to use new scientific knowledge, the firm with an offensive strategy aims to get there many years sooner. Even if it does not conduct oriented fundamental research itself it will need to be able to communicate with those who do, whether by the performance of applied research, through consultants or through recruitment of young postgraduates or by other means. This has very important implications for manpower policy as well as for communications with the outside scientific and technological community.

But although access to basic scientific knowledge may often be important, the most critical technological functions for the firm pursuing an offensive innovation strategy will be those centred on experimental development work. These will include design-engineering on the one hand, and applied research on the other. A firm wishing to be ahead of the world in the introduction of a new product or process must have a very strong problem-solving capacity in designing, building and testing prototypes and pilot plants. Its heaviest expenditures are likely to be in these areas, and it will probably seek patent protection not only for its original breakthrough inventions but also for a variety of secondary and follow-up inventions. Since many new products are essentially engineering 'systems', a wide range of skills may be needed. Pilkington's were successful with the 'float glass' process and IG Farben with PVC, largely because they had the scientific capacity to resolve the problems which cropped up in pilot plant work and could not be resolved by 'rule of thumb'. The same is even more true of nuclear-reactor development work.

There has been a great deal of confusion and misunderstanding over expenditure on R & D in relation to the total costs of innovation. It became fashionable to talk of R & D costs as a relatively insignificant part of the total costs of innovation —at most 10 per cent. This view is not supported by any empirical research and is based on a misreading of a United States Department of Commerce report frequently quoted and re-quoted. The small amount of empirical research which has been done on this question indicates that R & D costs typically account for about 50 per cent of the total costs of launching a new product in the electronics and chemical industries. As in so many aspects of industrial innovation it is Mansfield and his colleagues (1971 and 1977) who got down to the hard task of systematic empirical observation and measurement, rather than plucking generalizations from the air. Their results were confirmed on a larger scale by the Canadian surveys of industrial R & D and more recently by German work (OECD, 1982).

This is not to minimize the importance of production planning, tooling, market research, advertising and marketing. All of these functions must be efficiently performed by the innovating firm, but its most important distinguishing feature is likely to be its heavy commitment to applied research and experimental development. As we have seen, this was characteristic of IG Farben, Du Pont, GE, RCA, Bell and other offensive innovators. In the case of the new firm established to launch a new product, the inventor–entrepreneur is himself the living embodiment of this characteristic.

However, in order to succeed in its 'offensive' strategy the firm will not only need to be good at R & D, it will also need to be able to educate both its customers and its own personnel. At a later stage these functions may be socialized as the new technology becomes generally established, but in the early stages (which may last for some decades) the innovating firm may have to bear the brunt of this educational and training effort. This may involve running courses, writing manuals and textbooks, producing films, providing technical assistance and advisory services and developing new instruments. Typical examples of this aspect of innovation are the Marconi school for wireless operators, the BASF agricultural advisory stations, the ICI technical services for polyethylene and other plastics, the IBM and ICL computer training and advisory services, UKAEA's work on isotopes, and technical education of the consortia and the CEGB. As we have seen, many observers (e.g. Brock, 1975) believe that the efficient provision of these services was the decisive advantage of IBM in the world computer market.

The 'offensive' innovator will need good scientists, technologists and technicians for all these functions as well as for production and marketing of the new product.

This means that such firms are likely to be highly 'education-intensive' in the sense of having an above average ratio of scientifically trained people in relation to their total employment. The generation and processing of information occupy a high proportion of the labour force, but whereas for the 'traditional' firm this would represent a 'top heavy' and wasteful deployment of resources, these activities are the life-blood of the 'offensive' innovating firm.

'Defensive' innovation strategy

Only a small minority of firms in any country are willing to follow an 'offensive' innovation strategy, and even these are seldom able to do so consistently over a long period. Their very success with original innovations may lead them into a position where they are essentially resting on their laurels and consolidating an established position. They will in any case often have products at various stages of the product cycle—some completely new, others just established and still others nearing obsolescence. The vast majority of firms, including some of those who have once been 'offensive' innovators, will follow a different strategy: 'defensive', 'imitative', 'dependent', 'traditional', or 'opportunist'. It must be emphasized again that these categories are not pure forms but shade into one another. The differences assume particular importance in relation to industry in the developing countries, but they are important in Europe and America as well.

A 'defensive' strategy does not imply absence of R & D. On the contrary a 'defensive' policy may be just as research-intensive as an 'offensive' policy. The difference lies in the nature and timing of innovations. The 'defensive' innovators do not wish to be the first in the world, but neither do they wish to be left behind by the tide of technical change. They may not wish to incur the heavy risks of being the first to innovate and may imagine that they can profit from the mistakes of early innovators and from their opening up of the market. Alternatively, the 'defensive' innovator may lack the capacity for the more original types of innovation, and in particular the links with fundamental research. Or they may have particular strength and skills in production engineering and in marketing. Most probably the reasons for a 'defensive' strategy will be a mixture of these and similar factors. A 'defensive' strategy may sometimes be involuntary in the sense that a would-be 'offensive' innovator may be out-paced by a more successful offensive competitor.

Several surveys (Nelson, Peck and Kalachek, 1967; Schott, 1975 and 1976; Sirilli, 1982) have shown that in all the leading countries, most industrial R & D is 'defensive' or 'imitative' in character and concerned mainly with minor 'improvements', modifications of existing products and processes, technical services and other work with short time horizons. Defensive R & D is probably typical of most oligopolistic markets and is closely linked to product differentiation. For the oligopolist, defensive R & D is a form of insurance enabling the firm to react and adapt to the technical changes introduced by competitors. Since 'defensive' innovators do not wish to be left too far behind, they must be capable of moving rapidly once they decide that the time is ripe. If they wish to obtain or retain a significant share of the market they must design models at least as good as the early innovators and preferably incorporating some technical advances which differentiate their products, but at a lower cost. Consequently, experimental development and design are just as important for the 'defensive' innovator as for the 'offensive' innovator. Computer firms which continued to market valve designs long after the introduction of semiconductors could not survive. Chemical contractors which attempted to market a process which was technically obsolescent could not survive either. The 'defensive' innovator must be capable at least of catching up with the game, if not of 'leap-frogging'.

In an interesting sudy of the computer market, Hoffmann (1976) maintains that IBM has mainly followed a 'defensive' innovation strategy, although with some 'offensive' elements, while Sperry Rand (Univac) has pursued a more consistently 'offensive' strategy and Honeywell an 'imitative' strategy. Since IBM spends far more on R & D than Sperry Rand in absolute terms (Table 4.5a), this illustrates the point that the 'defensive' innovator may well commit greater scientific and technical resources than the 'offensive' innovator. A certain amount of 'slack' may be necessary in order to cover many new possibilities and to retain the flexibility needed to move very fast in catching up with the technical advances first introduced by competitors. (In interpreting the R & D figures in Table 4.5a, it must be remembered that Sperry Rand and Honeywell are also major producers of scientific instruments and that their strategies in these product fields may well differ.)

Patents may be extremely important for the 'defensive' innovator but they assume a slightly different role. Whereas for the pioneer patents are often a critical method of protecting a technical lead and retaining a monopolistic position, for the 'defensive' innovator they are a bargaining counter to weaken this monopoly. The defensive innovators will typically regard patents as a nuisance, but will claim that they have to get them to avoid being excluded from a new branch of technology. The offensive innovators will often regard them as a major source of licensing revenue, as well as protection for the price level needed to recoup R & D costs. They may fight major legal battles to establish and protect their patent position (RCA with television, ICI with polyethylene, La Roche with tranquillizers, Telefunken with PAL), and typically their receipts from licensing and know-how deals will far exceed expenditure. (In 1971, ICI had receipts of £13 million and expenditure of £3 million.)

The 'defensive' innovators will probably find it necessary to devote resources to the education and training of their customers as well as their own staff. They will also usually have to provide technical assistance and advice and these functions may be just as important for the 'defensive' as for the 'offensive' innovators. On the other hand, advertising and selling organizations, the traditional weapons of the oligopolist, will probably be more important, and to some extent technical service to customers will be bound up with this. The oligopolist may well attempt to use a combination of product differentiation and technical services to secure a market share not attainable by sheer originality (Brock, 1975; Hoffmann, 1976).

Both the 'offensive' and the 'defensive' innovator will be deeply concerned with long-range planning, whether or not they formalize this function within the firm. In many cases this may still often be the 'vision' of the entrepreneur and his immediate associates, but increasingly this function, too, is becoming professionalized and specialized, so that 'Product Planning' is a typical department for both 'offensive' and 'defensive' innovators. However, the more speculative type of 'technological forecasting' is more characteristic of the 'offensive' innovator, and as we have seen in chapter 7, still has considerable affinities to astrology or fortunetelling. It should probably still be regarded as a kind of sophisticated war dance to mobilize a faction in support of a particular project or strategy, but increasingly important serious techniques are being developed (Bright, 1968; Beattie and Reader, 1971; Encel et al., 1975; and Jones (ed.), 1981).

The 'defensive' innovator, then, like the 'offensive' innovator, will be a knowledge-intensive firm, employing a high proportion of scientific and technical manpower. Scientific and technical information services will be particularly important, and so will speed in decision-making, since survival and growth will depend to a considerable extent on timing. The defensive innovators can wait until they see how the market is going to develop and what mistakes the pioneers make, but they dare not wait too long or they may miss the boat altogether, or slip into a position of complete dependence in which they have lost their freedom of manoeuvre. R & D will be geared to speed and efficiency in development and design work, once management decides to take the plunge. Such firms will sometimes describe their R & D as 'advanced development' rather than 'research'.

Most commonly, the large multi-product chemical or electrical firm will contain elements of both 'offensive' and 'defensive' strategies in its various product lines, but a 'defensive' strategy is more characteristic of firms in the smaller industrialized countries, which cannot risk an 'offensive' strategy or lack the scientific environment and the market.

The strategy which a firm is able or willing to pursue is strongly influenced by its national environment and government policy. Thus, for example, European firms since the war have generally been unable or unwilling to attempt offensive innovations in the semiconductor industry and their role has been almost entirely 'defensive'. French chemical firms have followed a 'defensive' strategy while German chemical firms have often been 'offensive'. The complex interplay of national environment and firm strategy cannot be dealt with in detail here. But it is important to make the simple but fundamental point that many firms in the 'offensive' group are United States firms, while most firms in the developing countries are 'imitative', 'dependent' or 'traditional', with Europe in an intermediate position. This means that a 'defensive' innovation strategy has been particularly characteristic of European firms since the war. An over-simplified interpretation of Japanese experience since 1900 would be in terms of the movement of an increasing proportion of firms from traditional to imitative strategies, and then to defensive and offensive innovations. Japanese national policy has been designed to facilitate this progression. The extent of this shift is clearly observable in the statistics of the Japanese 'technological balance of payments' since the war. In the early post-war period Japanese firms were spending far more on buying foreign licences and know-how than they were themselves receiving from the sale of their own technology. At this time it was customary to regard the Japanese as 'superb imitators' and the long-term elements in their strategy were often overlooked. During the 1970s and 1980s on the *new* contracts which they have signed, Japanese firms receive more from the sale of their own technology than they pay out.

A technology policy of this sort involves a gradual change in the 'mix' of STS in the direction of a more R & D-intensive mix. The type of R & D also changes from adaptive to increased originality, but it may require a long period in which most enterprises follow a dependent or imitative strategy, whilst slowly strengthening their technical resources, on the basis of a carefully conceived long-term national policy, involving protection of 'infant technology' as well as the build-up of a wide range of government-supported STS. The main elements of this long-term national strategy have been well described by Allen (1981). The precise balance of STS must vary with the size, resource endowment and historical background of each country. But in many developing countries STINFO (Scientific and Technical Information Services), Survey organizations, Standard Institutes, Technical

Assistance organizations and Design–Engineering Consultancy organizations capable of impartial scrutiny and feasibility studies for projects involving imported technology are all of critical importance. They can provide the essential science and technology infrastructure which enables the STS at enterprise level to function effectively, despite the inevitable limitations in trained scientific and technical manpower. Only a few enterprises will gradually be able to develop first an adaptive and later an original innovative capacity. However, even in the United States the vast majority of firms are 'traditional', 'dependent' or 'imitative' in their strategies. We now turn to a consideration of these alternatives.

'Imitative' and 'dependent' strategies

The 'defensive' innovators do not normally aim to produce a 'carbon' copy imitation of the products introduced by early innovators. On the contrary, they hope to take advantage of early mistakes to improve upon the design, and they must have the technical strength to do so. At least they would like to differentiate their products by minor technical improvements. They will try to compete by establishing an independent patent position rather than simply by taking a licence, but if they do take a licence it will usually be with the aim of using it as a springboard to do better. However, their expenditure on acquisition of know-how and licences from other ('offensive' and 'defensive') firms may often exceed their income from licensing. For the 'imitative' firm it will always do so.

The 'imitative' firm does not aspire to 'leap-frogging' or even to 'keeping up with the game'. It is content to follow way behind the leaders in established technologies, often a long way behind. The extent of the lag will vary, depending upon the particular circumstances of the industry, the country and the firm. If the lag is long then it may be unnecessary to take a licence, but it still may be useful to buy know-how. If the lag is short, formal and deliberate licensing and know-how acquisition will often be necessary. The imitative firm may take out a few secondary patents but these will be a by-product of its activity rather than a central part of its strategy. Similarly, the imitative firm may devote some resources to technical services and training but these will be far less important than for the innovating firms, as the imitators will rely on the pioneering work of others or on the socialization of these activities, through the national education system. An exception to this generalization might be in a completely new area (e.g. in a developing country) when neither imports nor the subsidiary of an innovating firm have opened up the market. The enterprising 'imitator' may aspire to become a 'defensive innovator', especially in rapidly growing economies.

The 'imitator' must enjoy certain advantages to enter the market in competition with the established innovating firms. These may vary from a 'captive' market to decisive cost advantages. The 'captive' market may be within the firm itself or its satellites. For example, a large user of synthetic rubber, such as a tyre company, may decide to go into production on its own account. Or it may be in a geographical area where the firm enjoys special advantages, varying from a politically privileged position to tariff protection. (This will be the typical situation in many developing countries.) Alternatively or additionally, the imitator may enjoy advantages in lower labour costs, plant investment costs, energy supplies or material costs. The former are more important in electrical equipment, the latter in the chemical industry. Lower material costs may be the result of a natural advantage or of other activities (e.g. oil refineries in the plastics industry). Finally, imitators may enjoy advantages in managerial efficiency and in much lower overhead costs, arising from the fact that they do not need to spend heavily on R & D, patents, training, and technical services, which loom so large for the innovating firm. The extent to which imitators are able to erode the position of the early innovators through these advantages will depend upon the continuing pace of technological change. The early innovators will try to maintain a sufficient flow of improvements and new 'generations' of equipment, so as to lose the 'imitators'. But if the technology settles down, and the industry becomes 'mature', they are vulnerable and may have to innovate elsewhere. Du Pont's decision to move right out of the rayon industry despite their technical strength is a good example of strategic planning of this kind. Hirsch (1965) has summarized the characteristics of the product cycle which may permit 'imitators' to compete (Table 8.3 and Fig. 8.2). The extent to which they are actually able to do so, particularly in developing countries, is strongly influenced by institutional factors and government policies.

Unless the 'imitators' enjoy significant market protection or privilege they must rely on lower unit costs of production to make headway. This will usually mean that in addition to lower overheads, they will also strive to be more efficient in the basic production process. They may attempt this by process improvements, but both static and dynamic economies of scale will usually be operating to their competitive disadvantage, so that good 'adaptive' R & D must be closely linked to manufacturing. Consequently, production engineering and design are two technical functions in which the imitators must be strong. Even if they are making carbon copies under licence, the imitators cannot afford to have high production costs unless they have high tariff protection. They will also wish to be well-informed about changes in production techniques and in the market, so that scientific and technical information services are another function which is essential for the 'imitator' firm. The information function is also important for the selection of products to imitate and of firms from which to acquire know-how. It is clear that in all of this the would-be imitator in the typical developing country may be severely handicapped by local circumstances, unless national policies are carefully designed to facilitate technical progress.

A 'dependent' strategy involves the acceptance of an essentially satellite or subordinate role in relation to other stronger firms. The 'dependent' firm does not attempt to initiate or even imitate technical changes in its product, except as a result of specific requests from its customers or its parent. It will usually rely on its customers to supply the technical specification for the new product, and technical advice in introducing it. Most large firms in industrialized countries have a number of such satellite firms around them supplying components, or doing contract fabrication and machining, or supplying a variety of services. The 'dependent' firm is often a sub-contractor or even a sub-sub-contractor. Typically, it has lost all initiative in product design and has no R & D facilities. The 'small' firms in capital-intensive industries are often in this category and hence account for hardly any innovations (see chapter 6). For the special role of such firms in the Japanese economy, see Clark (1979).

The pure 'dependent' firm is in effect a department or shop of a larger firm, and very often such firms are actually taken over. But it may suit the large firm to maintain the client relationship, as sub-contractors are a useful 'cushion' to mitigate fluctuations in the work load of the main firm. The 'dependent' firm may also wish to retain its formal independence as the owners may hope they will ultimately be able to change their status by diversification or by enlarging

Table 8.3   Characteristics of the product cycle

Characteristics	Cycle phase		
	Early	Growth	Mature
Technology	short runs rapidly changing techniques dependence on external economies	mass production methods gradually introduced variations in techniques still frequent	long-runs and stable technology few innovations of importance
Capital intensity	low	high, due to high obsolescence rate	high, due to large quantity of specialized equipment
Industry structure	entry is know-how determined numerous firms providing specialized services	growing number of firms many casualties and mergers growing vertical integration	financial resources critical for entry number of firms declining
Critical human inputs	scientific and engineering	management	unskilled and semi-skilled labour
Demand structure	sellers' market performance and price of substitutes determine buyers' expectations	individual producers face growing price elasticity intra-industry competition reduces prices product information spreading	buyers' market information easily available

Source: Hirsch (1965).

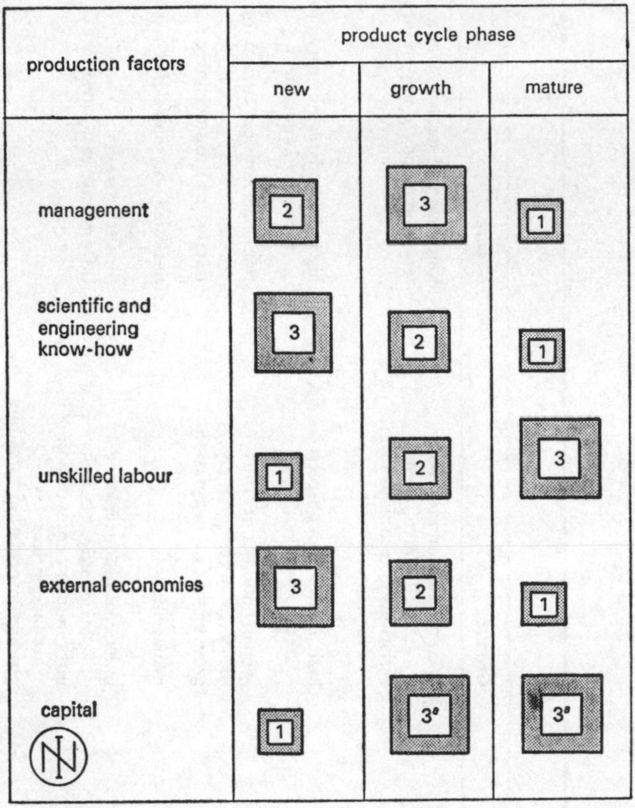

production factors	product cycle phase		
	new	growth	mature
management	2	3	1
scientific and engineering know-how	3	2	1
unskilled labour	1	2	3
external economies	3	2	1
capital	1	3[a]	3[a]

Fig. 8.1   The relative importance of various factors in different phases of the product cycle.
The purpose of the blocks is simply to rank the importance of the different factors, at different stages of the product cycle. The relative areas of the rectangles are not intended to imply anything more precise than this.[a] Considered to be of equal importance.
Source: Hirsch (1965).

their market. They may in any case prize even that limited degree of autonomy which they still enjoy as a satellite firm. In spite of their apparently weak bargaining position, they may enjoy good profits for considerable periods, because of low overheads, entrepreneurial skill, specialized craft knowledge or other peculiar local advantages. Even if they are 'squeezed' pretty hard by their customers, they may prefer to endure long periods of low profitability rather than be taken over completely. Although bankruptcies and take-overs may be common, there is also a stream of new entries.

'Traditional' and 'opportunist' strategies

The 'dependent' firm differs from the 'traditional' in the nature of its product. The product supplied by the 'traditional' firm changes little, if at all. The product supplied by the 'dependent' firm may change quite a lot, but in response to an initiative and a specification from outside. The 'traditional' firm sees no reason to change its product because the market does not demand a change, and the competition does not compel it to do so. Both lack the scientific and technical capacity to initiate product changes of a far-reaching character, but the 'traditional' firm may be able to cope with design changes which are essentially fashion rather than technique. Sometimes indeed, this is its greatest strength.

'Traditional' firms may operate under severely competitive conditions approximating to the 'perfect competition' model of economists, or they may operate under conditions of fragmented local monopoly based on poor communications, lack of a developed market economy, and pre-capitalist social systems. Their technology is often based on craft skills and their scientific inputs are minimal or non-existent. Demand for the products of such firms may often be very strong, to some extent just *because* of their traditional craft skills (handicrafts, restaurants and decorators). Such firms may have good survival power even in highly industrialized capitalist economies. But in many branches of industry they have proved vulnerable to exogenous technical change. Incapable of initiating technical innovation in their product line, or of defensive response to the technical changes introduced by others, they have been gradually driven out. These are the 'peasants' of industry.

An industrialized capitalist society includes some industries which are predominantly 'traditional', and others characterized by rapid technical innovation. It has been argued that an important feature of the twentieth century has been the growth of the 'research-intensive' sector. But it is a matter of conjecture and of policy as to how far this change may continue. It is a complex process, since sometimes the very success of a technical innovation may lead to standardized mass production of a new commodity with little further technical change or research for a long time. Usually, however, the industries generated by R & D have continued to perform it, so that the balance has gradually shifted towards a more research-intensive economy, and a higher rate of technical change. It is the contention of this book that this is one of the most important changes in twentieth-century industry, but it must be seen over a long time perspective.

This shift has been less the result of any conscious central government strategy (although government policies have increasingly tended to favour this change) than the outcome of a long series of adaptive responses by firms to external pressures at home and abroad, and of attempts to realize the dreams of inventors. The efforts of firms to survive, to make profits and to grow have led them to adopt one or more of the strategies which have been discussed. But the variety of possible responses to changing circumstances is very great, and to allow for this element of variety one other category should be included, which may be described as an 'opportunist' or 'niche' strategy. There is always the possibility that entrepreneurs will identify some new opportunity in the rapidly changing market, which may not require any in-house R & D, or complex design, but will enable them to prosper by finding an important 'niche', and providing a product or service which consumers need, but nobody else has thought to provide. Imaginative entrepreneurship is still such a scarce resource that it will constantly find new opportunities, which may bear little relation to R & D, even in 'research-intensive' industries.

Innovation strategy in developing countries

Those firms which adopt a strategy of offensive or defensive innovation have gradually 'learned' how to innovate. But there is no recipe which can ensure success and intense controversy still surrounds the important ingredients. The fact that they are often innovating on a world market increases the uncertainty which they confront, and has led increasingly to the involvement of government to subsidize R & D, to create appropriate infrastructures and to diminish market uncertainty. Economic policy inevitably becomes enmeshed with policy for science and technology. These problems are particularly acute for the developing countries.

An underdeveloped economy may for a while base itself mainly or entirely on an industrial structure which relies on dependent and traditional strategies. If it does so, it is likely to remain extremely poor and backward. One possible alternative is the Chinese path but this is difficult for smaller and weaker nations. Even a successful imitative strategy, although it may lead to industrial development, will reach a point when export competitiveness in labour costs may increasingly conflict with the goal of higher *per capita* incomes. In such a case, the Japanese strategy of moving up the scale steadily may be the most appropriate, and the distinctive feature of the Japanese success has been the way in which government policies have underpinned the efforts of management at enterprise level (Allen, 1981 and Peck and Wilson, 1982). However, the Japanese success in raising *per capita* incomes rapidly and in strengthening the technical capacity of the economy, was accompanied by considerable degradation of the environment and other unpleasant consequences of rapid industrial change. It is to these problems of national and international policy for innovation and for science and technology that we turn in the final chapter, but here it will be useful to consider very briefly some of the problems of developing countries in the context of the analysis in this chapter.

Innovative effort which is directed towards satisfying the market needs of consumers tends to be biased towards higher income groups for several reasons. Most obviously, of course, poorer people cannot afford much more than the bare necessities and they cannot afford to pay the premium prices which are often inevitable in the early stages of a new product. In the jargon of economics this means that new products tend to have a high income-elasticity. Wealthier people and richer firms can afford to indulge new tastes and to take more risks.

It is on a global scale that the most extreme effects of worldwide inequality in incomes are apparent. The bias in the world research innovation system is so great as to constitute a danger to the future of human society. The elementary facts are by now universally known. The Lorenz curve of world income distribution shows a skewness far more extreme than that of any individual country. Not so well known is the fact that over 90 per cent of the world's R & D is conducted in the industrialized countries, and that it is overwhelmingly and quite naturally directed towards satisfying demands in those countries (United Nations, 1970; Freeman, Cooper and Pavitt, 1978; Herrera, 1981). This means that very little of the world's R & D is in fact directly concerned with the elementary needs of the majority of the world's inhabitants. And here the bias in the *capital* goods sector is often of the greatest importance. The need for innovations in both capital goods and consumer goods designed specifically for the needs of the developing countries is very great; yet the innovation mechanism of the world market is biased overwhelmingly towards the high income countries. The bias is so strong that some non-US companies now actually launch their innovations first on the US market. The need for labour-intensive innovations can never be met in this way, and the need for new policies is urgent. The indiscriminate import of technologies developed for entirely different markets through the operations of multi-national corporations may have disastrous employment and other social effects in weak poor countries (Cooper, 1973).

The import of foreign technology is often discussed in terms of two equally impracticable extremes. On the one hand, a position of complete autarchy in science and technology, of striving to be completely independent in every single branch of research and development, would be ruinously expensive and almost impossible for all but the largest super-powers. The mechanisms for the international transfer of technology are of the greatest importance for policy-makers in the developing countries. Every country stands to gain enormously from international interchange and division of labour in world science and technology. On the other hand, an international division of labour in science and technology which is so one-sided that it leaves large areas virtually denuded of independent scientific capacity is equally unacceptable. Even on the narrowest economic grounds it is highly inefficient, and is only recommended by those economists who have had no practical contact with the problems of technology transfer. Simply to assimilate any sophisticated technology today, and operate it efficiently, requires *some* independent capacity for R & D, even if this is mainly adaptive R & D. Not just in agriculture, but also in manufacturing, the variety of local conditions is so great that simple 'copying' is often ruled out. Thus in many countries the capacity to receive technology from outside imperatively requires some independent indigenous science base. To solve the innumerable local problems of soil, materials, environment, skills and climate requires that the indigenous base should grow and flourish.

What is desirable on economic grounds is even more so on cultural and political grounds. While *some* scientific and technical capacity is necessary for assimilation of the results of foreign research and technical progress, it is undoubtedly possible to get by with a far smaller commitment than that made in the super-powers or even in several West European countries. Obviously the *size* of a country has a very great bearing on this question and will affect the degree of specialization which is necessary. Heavy reliance on imported technology is an inescapable necessity for most countries in the world. The economic consequences of this situation are perhaps not too serious, but the political and cultural consequences are very great. One must therefore expect that the smaller countries, as well as the developing countries, will lay increasing stress on equitable international arrangements for access to world science and technology. The attempt to establish more expensive 'autarchic' R & D is to some extent a defensive reaction against the political dangers of potential lack of access. Only in proportion to the growth of mutual trust, and a genuinely international policy, will the achievement of more equitable and mutually beneficial international division of labour in science and technology be possible. Such a division must in any case be based in principle on all countries *contributing* to as well as *drawing* from the world stock of knowledge.

The implications of this are complex for technology and multinational corporations but relatively clear for fundamental science. The greatest significance of fundamental research is that it provides a multi-purpose general knowledge base on which to build a wide range of scientific and technical services. Every country without exception requires such a base, even if only on a very small scale. Without it there cannot be any independent long-term cultural, economic or political development. One of the main objectives of world policy for science and technology should be to build and sustain an indigenous scientific capacity throughout the developing world. The Canadian International Development Research Centre was an important step towards the reorientation of world science in this direction (IDRC, 1972).

Conclusions

In Part One of this book it was argued from historical evidence that the professionalization of the R & D process was one of the most important social changes in twentieth-century industry. In Part Two it has been argued that the requirements of successful innovation and the emergence of an R & D establishment within industry have profoundly modified patterns of firm behaviour. This means that it is no longer satisfactory (if it ever was) to explain firm behaviour exclusively in terms of response to price 'signals' in an external environment, and adjustment towards an 'equilibrium' situation. World technology is just as much a part of the firm's environment as the world market, and the firm's adaptive responses to changes in technology cannot be reduced to predictable reactions to price changes. This makes things difficult for economists. It means that they must pay much more attention to engineers and to sociology, psychology and political science. Economists have an elegant theory which is confronted with a very untidy and messy reality. Their theory was and is an important contribution to the explanation and prediction of many aspects of firm behaviour, but it is not self-sufficient and attempts to make it so can only lead to sterility.

The sketchy discussion in this chapter is not intended as an alternative theory of firm behaviour. Such a theory requires a greater integrative effort in the social sciences. But it is intended to indicate the kind of issues which must be embraced by any theory which seeks to explain the firm's adaptive response to technological change, as well as to price changes in its factor inputs and the market for its products. There are encouraging indications that social scientists from several disciplines, including economists, are beginning to tackle the development of a more comprehensive and satisfactory theory of the firm. Particularly notable is the work of Mansfield (1968a and b; *et al.*, 1971 and 1977), Nelson (1971, 1977, 1980 and 1982) and Gold (1971 and 1979) in the United States, who have made outstanding empirical studies of firm behaviour in relation to innovation. Nelson's new work with Winter may at last bridge the chasm which has developed between the empirical findings discussed here and macroeconomic theory.

Much better known, of course, is the work of Galbraith (1969), who has shown great awareness of the relevance of technological innovation for economic theory. His emphasis on the increased specialization and complexity of technology and the emergence of a 'techno-structure' is fully consistent with the argument of Part One, but there are some important differences of interpretation which are discussed in the next chapter.

PART THREE

Innovation and Government

A society that blindly accepts the decisions of experts is a sick society on its way to death. The time has come when we must produce, alongside specialists, another class of scholars and citizens who have broad familiarity with the facts, methods and objectives of science and thus are capable of making judgements about scientific policies. Persons who work at the interface of science and society have become essential simply because almost everything that happens in society is influenced by science.

Dubos (1970, p. 227)

Man cannot predict the future, but he can invent it.

Dennis Gabor (1964)

The evidence in Part One of this book related mainly to a few research-intensive industries, especially chemicals and electronics. Although some additional empirical data has been cited in Part Two, it remains true that several important sectors of industry have been largely ignored. These fall into two main groups:

1. Other research-intensive industries, such as aircraft and nuclear weapons, where there has been a very heavy government involvement, both in procurement and in R & D itself.
2. Many consumer-goods industries and services mostly of low research-intensity, such as clothing, furniture, food, consumer durables and construction.

Much of the argument of the book is relevant to all sectors of industry. For example, the discussion on size of firm in chapter 6 and methods of project evaluation in chapter 7 is as relevant to the steel industry or to the food industry as it is to electronics and chemicals. But there are some features of these two groups of industries which raise new issues which have as yet hardly been discussed. These are primarily issues of public policy.

In the US, UK and France the aircraft industry accounted for more than a quarter of total industrial R & D expenditure during most of the post-war period (Table 1.3). By far the greater part of this was financed with public money, although the development work was carried out in industry. The recognition in the 1960s that it was no longer necessary or desirable to give such a high priority to the development of new types of *military* aircraft was often followed by increased programmes of public expenditure for *civil* aircraft development. In some cases this was no doubt justifiable in terms of civil, economic and social benefits, particularly for example in reducing the cost of air travel and some of the disamenity effects, such as noise. But there was little evidence that the Supersonic Transport (SST) could be justified in economic or welfare terms; nor was there any readiness by the manufacturers to finance the main part of the development costs. No private entrepreneurs would accept this risk, but a very high proportion of public expenditure on civil R & D in UK and France went to the Concorde and other aircraft projects in the 1960s. Only in the 1970s did this tail off with the completion of Concorde.

This is not to say that the *principle* of public support for civil, industrial R & D including aircraft development is wrong. Indeed it can be justified on a small scale on economic grounds, but *a priori* there is no more reason why such support should go to the aircraft industry rather than to the railways, telecommunications, the computer industry or machine tools. If a balance is struck between the social costs and benefits involved, all of these might be stronger candidates, even taking into account technological spin-off to other industries. Indeed, the spin-off effects of advanced land transport systems could quite possibly be greater than for air transport, as well as the direct benefits to consumers. The conclusion is difficult to escape that the preferential treatment of R & D in this industry was due less to any considered assessment of transport or communication needs than to habit, the continuing power of a lobby and the prestige elements associated. It was certainly not due to any sophisticated project evaluation techniques.

So far as the American experience is concerned, Nelson and Eads (1971) have argued that the attempt to emulate military 'crash' programmes in such civil techno-logies as the SST and nuclear reactors was misconceived and a considerable waste of resources was the result. Government expenditures, although very important, should have concentrated on applied research and early experimental development. It is in the area of fundamental research and enabling technologies that the economic case for public finance and public laboratories is overwhelming. When it comes to those stages of development directly linked to the introduction of commercial products or systems, it is much more likely that waste will be avoided if this is carried out by the enterprise and largely or entirely at the firm's expense. It is clear that there are very grave dangers in major government subsidies to firms to cover their development costs. If government subsidies to enterprises are used at all, then thorough public discussion of priorities is essential.

Whether or not one accepts the basic theme of Galbraith's *New Industrial State*, or the arguments advanced in Parts One and Two of this book, it would be difficult to deny that the 'military–industrial complex' is a reality which very much affects firm behaviour at least in a few industries. The scale and complexity of modern technology have been carried to extreme limits in research, design and development for military aircraft, missiles and nuclear weapons. The large-scale participation of governments and the peculiar nature of the military market mean that the process of 'advocacy' in project selection, which as we have seen is present in all R & D policy making, becomes overtly political at the national level. The 'lobby' and the 'corridor-padder' are more important in this type of decision-making than elaborate calculations of return on investment. Indeed such calculations may often be used purely as a gimmick to provide a pseudo-rationalistic method of manipulating the political process. Clearly national policies of a non-economic nature have predominated in determining the innovative performance of the aircraft industry, both military and civil, and the same is true for several other industries closely linked to aircraft (Peck and Scherer, 1962; Peck, 1968).

So far most of this book has been concerned with innovation at the level of the firm. This is because in capitalist societies most industrial R & D is performed by enterprises and innovations are made by firms. However, it has been apparent throughout, but especially from chapter 8, that governments carry an increasing responsibility for the overall framework within which firms attempt their innovations. They are also responsible for the expenditure of a large part of R & D funds in most countries, and for the performance of R & D in the government sector. The Second World War greatly increased their involvement, and the limitations of *'laissez-innover'* mean that public intervention is unlikely to diminish very much. Attempts at disengagement have not been conspicuously successful, as, for example, in the UK in the early 1970s and again in the early 1980s. Those countries like Japan, which have followed a more consistent and patient public policy towards technology, have on the whole been more successful, even if the scale of their government expenditures has been much smaller, because of the insignificance of military R & D.

Priorities in public R & D expenditures

This chapter discusses the priorities of government policies for science and techno-logy since the Second World War and suggests some changes in these priorities. A general tendency in most OECD countries has been towards a more explicit recognition of these priorities (Tisdell, 1981). But there is still considerable unwillingness to think in these terms. The philosophy expressed in the 'Rothschild Report' (1971) is still strong. According to this report, which was accepted and implemented by British governments during the 1970s, *national* priorities for R & D cannot ever be established. This can take place only at the departmental level, where each government department must behave as a 'customer' contracting for the R & D which it requires and acting autonomously.

As the example of the aircraft industry has shown, both in Europe and the US, *actual* priorities are established even if they are implicit rather than explicit. These real priorities can be recognized by examination of the actual distribution of R & D expenditures. One of the most characteristic features of the pattern of public expenditures in the 1950s and 1960s was the massive scale of nuclear, military and space programmes. This was particularly true of the US, UK and France, within the OECD, and of the Soviet Union and China outside. But it was also true in lesser measure of several other countries. Even in the German Federal Republic, which did not have any significant military programme in the 1950s, these programmes grew very rapidly so that they accounted for over a third of all public R & D expenditures by 1970. Typically, these heavy expenditures in the OECD countries were divided between 'intra-mural expenditures' allocated entirely to the governments' own R & D establishments, such as Farnborough or Aldermaston, and 'extra-mural' contract R & D expenditures in private industry.

Only in smaller countries and Japan did these outlays fall below 25 per cent of publicly financed R & D (Table 9.1), although Sweden had heavy programmes in

Table 9.1   The change of percentage shares of military, space and nuclear
R & D expenditures as a proportion of total public R & D
expenditure during the 1960s

Country	*1960–1*				*1969–70*			
	Defence	*Space*	*Nuclear*	*Total*	*Defence*	*Space*	*Nuclear*	*Total*
US	68.7	9.1	10.7	88.5	48.7	23.2	6.5	78.4
Canada	23.2	–	21.2	44.4	11.2	1.4	19.5	32.1
Belgium	6.0	–	24.3	30.3	2.0	6.0	14.8	22.8
UK	64.5	0.5	14.7	79.7	40.4	3.7	11.5	55.6
Norway	8.6	0.4	16.5	25.5	7.1	1.2	8.3	16.6
Japan	5.6	–	7.6	13.2	2.2	0.7	7.4	10.3
Sweden	49.0	0.1	23.9	73.0	28.3	1.5	9.4	39.2
Netherlands	5.0	0.2	11.7	16.9	4.5	2.9	10.5	17.9
France	41.5	–	27.5	69.0	30.7	6.7	17.8	55.2

Source: OECD Statistics (1971).   – nil

military and nuclear fields, accounting still for 60 per cent of public R & D expenditures in 1967. Since the US and the medium-sized European countries account for a very high proportion of total R & D in the OECD area, the proportion of all *public* R & ·D expenditures going to national security and prestige types of R & D was well over 75 per cent in the early 1960s. This was equivalent to nearly half of R & D expenditure of all kinds (public and private) in the OECD area in the 1960s. During the 1970s the pattern changed considerably (Table 9.2) for a variety of reasons. Already during the 1960s the US government's expenditure on NASA was falling after the successful moon landings, and the policies of international détente also facilitated some reduction in the relative scale of military expenditures. In Britain, however, the very high level of military spending was still further increased in the 1970s.

A classification of R & D expenditures by public goals must always be attended by great difficulties, since the same programme may be supported by different agencies for a variety of motives and, moreover, these motives are changing over time. To attempt such classification is nevertheless highly desirable, since the debate on public priorities and the desirable direction of change must be informed by knowledge of the approximate scale of allocations and the extent of the change which is contemplated or feasible. The classification which is followed here is that generally adopted by the OECD in its statistical publications on R & D. The early OECD statistics grouped together military, space and nuclear expenditures under the heading 'national security and prestige'. These categories were also often loosely referred to as 'big science' and 'big technology'. Although there are advantages in this type of classification, one must nevertheless recognize at the outset that it has certain limitations and the OECD has since improved upon it.

While most weapons-development expenditure can be assigned to the military heading with few complications, the problem is not so simple in the case of nuclear and space expenditures. In the early days of the nuclear programmes, military considerations usually predominated, and even the expenditures on basic research (which were extensive) were often voted under a defence heading. As the commercial development of nuclear power grew more important, and stockpiles of nuclear weapons were accumulated, the balance changed. Programmes which had at one time found support mainly for security and prestige reasons were now justifiable to a far greater extent for economic reasons and for the advance of science. It is now often possible to disaggregate such headings as 'nuclear' or 'space' expenditures and reclassify them to headings such as 'energy' or 'fundamental research'. Countries still differ, however, in the extent to which they are willing and able to make this reclassification and in intepreting Table 9.2 these difficulties must be kept in mind. Nevertheless, these problems of classification do not affect the main picture emerging from Table 9.2—the relatively low priority given to environmental and welfare research by comparison with military R & D, especially in the UK, US and France. Nevertheless, resources devoted to R & D for environmental protection, as well as energy, were growing in the 1970s.

There is an important sense in which the grouping together of military, nuclear and space programmes, at any rate in the 1950s and 1960s, may be justified. All of them depended almost exclusively on public funds during this period; all of them were heavily if not exclusively influenced by considerations of national security and prestige; and all of them involved special institutions to control and operate the programmes, usually of the 'big' variety. Moreover, they became identified in the public mind with a definite kind of science policy. The development of this public 'image' of 'big science and technology' has had very important social consequences. The priority accorded to these projects was so great that not without some justification, sections of public opinion have tended to accept 'big science and technology' as the image of science and technology in general.

Changing values and changing priorities

Since the Second World War, policies for science and technology in the OECD countries have gone through several different phases. In the immediate post-war

Table 9.2 Total specific government R & D funding by socio-economic objective (percentage distribution)

	United States 1971	1975	1980	Japan[a] 1975	1979	Germany 1971[b]	1975	1980	United Kingdom 1971[b]	1975	1980	France 1971[b,c]	1975	1980
Defence	52.2	50.8	47.0	3.8	3.6	21.3	17.6	14.2	46.2	52.9	59.4	38.0	32.6	40.9
Space	19.2	14.5	14.4	11.8	9.3	9.4	6.8	6.0	1.9	2.5	2.3	7.0	6.1	5.0
Civil aeronautics	3.1	1.6	1.6	—	—	3.6	2.6	2.3	14.5	8.2	3.4	7.0	6.7	2.4
Defence and aerospace	74.9	66.9	63.0	15.6	12.9	34.3	27.0	22.5	62.6	63.6	65.1	52.0	45.4	48.3
Agriculture	1.9	2.2	2.2	22.2	18.4	3.1	3.0	2.6	2.9	4.8	4.5	4.0	4.2	4.3
Industrial growth n.e.c	0.6	0.4	0.4	17.7	13.9	8.6	9.1	11.7	4.6	3.1	3.4	7.0	8.9	7.6
Agriculture and industry	2.5	2.6	2.6	39.9	32.3	11.7	12.1	14.3	7.5	7.9	7.9	11.0	13.1	11.9
Production of energy	3.6	7.1	11.8	12.8	17.8	16.4	16.8	20.1	7.5	7.1	7.3	8.0	9.4	8.5
Transport, telecommunications	1.6	1.8	1.1	3.2	2.2	0.9	2.3	2.9	0.9	0.7	0.7	⎫	3.2	3.2
Urban and rural planning	0.4	0.5	0.4	1.0	1.9	0.8	1.8	2.1	1.2	1.7	1.1	⎬ 6.0	1.6	1.5
Earth and atmosphere	1.5	2.0	2.0	1.4	1.9	2.3	2.8	3.9	0.3	0.8	0.9	⎭	3.3	3.3
Energy and infrastructure	7.1	11.4	15.3	18.4	23.8	20.4	23.7	29.1	9.9	10.3	10.1	15.0	17.5	16.5
Environment protection	0.9	0.9	1.1	2.6	2.5	0.5	1.6	2.8	0.2	0.6	0.9	⎱ 3.0	0.9	1.2
Health	8.7	11.9	11.9	5.1	4.3	4.4	5.2	5.7	1.9	2.3	1.8	⎰	4.4	5.0
Social development and services	2.6	2.0	2.2	2.0	1.5	6.7	7.7	5.4	0.7	1.2	1.2	1.0	1.2	1.4
Health and welfare	12.2	14.8	15.2	9.7	8.3	11.6	14.5	13.9	2.8	4.1	3.9	4.0	6.5	7.6
Advancement of knowledge n.e.c[d]	3.3	4.3	3.9	2.8 / 13.6[e]	2.5 / 20.2[e]	22.0	22.7	20.2	17.2	14.1	13.0	19.0	17.0	15.2
Total specified R & D funding[d]	100.0	100.0	100.0	100.0	100.0	100.0	100.0	100.0	100.0	100.0	100.0	100.0	100.0	100.0

[a] Government intramural only, except for Advancement of Knowledge and Industrial Development.
[b] Not strictly comparable with following years.
[c] Rough OECD estimate.
[d] Excludes public general university funds throughout and also excludes basic research supported by US mission-oriented agencies. An 'adjusted' US figure might be about 15 per cent in 1980.
[e] Total university receipts from government for specified projects including those for other objectives.

n.e.c = not elsewhere classified.

Source: OECD (1981).

period attitudes were still heavily influenced by the experiences of the war itself and of course by the beginning of the Cold War. During this first period the emphasis was very heavily on the 'supply side' of the science/technology system, and especially on building up strong R & D capability. In the 1960s and 1970s, this gave way increasingly to a more balanced approach which recognized the dangers of a one-sided reliance on R & D policy and put greater emphasis on the general economic environment affecting technical change and on the innovation process as a whole. In a crude and over-simplified way this second phase might be described as a 'demand' period by contrast with the 'supply' orientation of the late 1940s and 1950s.

Finally, in the most recent years, there are increasing attempts to integrate both these approaches and to link up policies for science and technology with policies for industry and for the economy generally. These differences of emphasis should not be exaggerated; elements of the 'supply' and the 'environment' approaches were of course present in most countries all the way through.

The tremendous success of the Manhattan Project (the development of the atomic bomb), the radar programme and military aircraft during the Second World War had convinced governments that an enormous investment in science and technology could produce an astonishing payoff in purely military terms. They appeared, at least superficially, to justify the view that the assembly of large R & D teams with generous financial support could solve many difficult and complex problems. The effectiveness of weapons had been increased, not by a small percentage, but by orders of magnitude. Consequently, as the tensions of the Cold War increased, there was a readiness to invest even vaster sums in the development of the H-bomb and appropriate delivery systems. It should be noted, as Hitch has pointed out (1962), that the military establishment was by no means always so research-minded and was once a by-word for hidebound opposition to science. As he has suggested, this may have major implications for other civil professional groups who have not yet come to accept the revolutionary potentiality of a big investment in modern science and technology. The thought is often expressed that if human beings can use science and technology to get to the moon, it should not be beyond the wit of man to solve some of our more urgent terrestrial problems, such as urban transport. Although naïvely formulated, this thought nevertheless contains an important truth. What science and technology can achieve *is* partly a question of the social priorities and goals set for research (Nelson, 1977).

One of the consequences of the successful development of the A-bomb was to give immense prestige and weight to the nuclear and aircraft lobbies[1] in national decision-making for R & D. The fashion set in the 1940s dominated world R & D expenditures in the next decade. In the early 1950s at the height of the Cold War, the powers which then led the nuclear weapons race were devoting more than half their national R & D resources and more than 90 per cent of their public expenditure to these objectives. It should be remembered that the R & D programmes were effective in the narrow sense in which they were conceived. Although attended by fantastic cost overruns, they nevertheless produced successive generations of ever more sophisticated weaponry. Moreover, those who queried the extraordinarily high priority given to these programmes were often met with the 'spin-off' argument according to which they benefited technology and economic

---

[1] The expression 'lobby' is used here, and throughout the book, not in a pejorative sense, but in the normal sense of political science, to describe an interest group with a distinctive set of attitudes.

progress in general. The demonstration effect of the Manhattan Project not only served to justify a succession of 'big technology' projects in the countries which originated it, but also led to a wave of competitive and imitative efforts. Its repercussions were not confined to the military field, but set the tone in national priorities in fundamental research and civil technology (see, for example, Greenberg, 1969). The primacy of nuclear physics in civil research, and of nuclear energy programmes in national fuel research, cannot be entirely dissociated from the socio-political consequences of the awesome achievements of nuclear weapons. The first *de facto* science policy institutions in many countries were government nuclear research organizations (Dedijer, 1964).

The launching of Sputnik in 1957 had perhaps even greater repercussions. Although it probably owed its successful development to the overall priority given to rocket-type delivery systems for nuclear weapons, its importance obviously transcended these objectives. It led immediately to a further massive competitive increase in American public expenditures on R & D, and to major changes in policy for science and technology in many other countries. It impelled the American President to set as the major national priority for US technology the prestige objective of putting a man on the moon by 1970. This objective was triumphantly achieved, and few would question the magnitude of the technical achievement.

An over-emphasis on 'big science and technology' was not the only weakness of the science policies of the 1950s. There was a general over-estimation of the importance of R & D and a relative neglect of other scientific and technical services, and of those other activities which are essential for successful innovation and efficient technical change throughout the economy. With a few exceptions, such as Nelson, Carter and Williams, economists had largely neglected these topics and knew little or nothing about R & D or innovation. Among leading economists only Schumpeter had given invention and innovation pride of place in his models of the behaviour of the economic system and even he took little interest in the policy implications of his analysis, whether for government or industry. A typical pre-war economics textbook (and many of the post-war ones) had nothing to say about R & D or even innovation, and it was not until the 1960s that most economists began to take these questions more seriously. There were very few studies of innovation and when he surveyed the literature, Rogers (1962) could find only one piece of empirical research on the diffusion of innovation in industry by an industrial economist. Keynesian macro-economics dominated teaching and research and the general assumption tended to be that if economic policy could maintain a high level of effective demand, technical change was a black box which need not be opened.

This did not mean, however, that economists were unmindful of the claims of science. On the contrary, they were generally quite ready to treat science and technology with great generosity. Formally speaking, economic theory of almost all kinds recognized that technical progress was the mainspring of economic growth, even though it showed little disposition to delve into its mysteries. As we have seen, from Adam Smith onwards all the great economists had treated science and invention as worthy of special government blessing and promotion. Only Milton Friedman has recently broken with this tradition and much more typical of neo-classical economics was Kenneth Arrow's (1962) sophisticated demonstration that a market economy would tend to under-invest in research. That this deficiency should be compensated by government funding was common ground and virtually conventional wisdom, even for the majority of monetarist iconoclasts.

The benign neglect of the economists, the ignorance of the politicians and many managers, and the self-interest of the R & D and military establishments combined to create favourable conditions for the extraordinarily rapid growth of expenditures on R & D, both by government and industry. The 1940s and 1950s were the golden age of R & D expansion when very few questions were asked about the efficient use of funds and the general assumption was that increased expenditure could only do good. This was not quite so naïve as it may sound today as the evidence of a very high rate of return on R & D investment was quite strong both in the military and the civil spheres (Schott, 1975). Arrow's (1962) and Nelson's (1959) arguments about the tendency of a market economy to under-invest were relevant in many other areas besides basic research, so that even with the benefit of hindsight a fairly rapid expansion of many R & D budgets had a good economic justification.

It was a biologist, Julian Huxley (1934), and a physicist, J. D. Bernal (1939), and not economists, who had made the first systematic attempt to measure the resources devoted to R & D in a national market economy in the 1930s. In his book, *The Social Function of Science*, J. D. Bernal estimated that the total British expenditures at that time were about 0.1 per cent of national income. This was almost certainly a substantial underestimate because of the inadequate data on the D part of industrial R & D. But if we allow for a revised figure of 0.2 or 0.3 per cent, then there was still more than an order of magnitude increase in the real resources devoted to R & D over the next thirty years. This was indeed what Bernal himself had advocated, but as a Marxist critic of an ailing capitalist economy, he scarcely expected that his proposals would be adopted in many capitalist countries as well as in the socialist countries.

Most economists have given up now on the purely statistical attempts to disaggregate the aggregate production function and the disaggregation of the components of 'technical change'. Few would accept Denison's (1962) heroic two-decimal place estimates of the 'contribution' of R & D or education to the growth of the US economy as at all accurate. Nevertheless, a plausible case can be made for the view that one of the important reasons for the world-wide economic growth rate in the 1950s and 1960s being higher than in any previous quarter century was the high rate of technical change sustained by the massive expansion of R & D, especially in manufacturing industry. Even the high opportunity cost of the huge military, nuclear and space programmes in some countries did not prevent the achievement of great economic benefits from the R & D system as a whole, as well as from the world-wide diffusion of technology.

The second stage of policies for science and technology

Our evaluation of the first period of headlong expansion after the war should by no means be purely negative. But it was bound to give way to a period of slower growth or even contraction and to a method of managing R & D and other technical activities which was far more cost-conscious and far more concerned with cost-effectiveness. There were many reasons for this transition and generally industry was ahead of government in making it, for obvious reasons. The pressures of the market place, of competition and of profitability compelled many firms to review their R & D projects and programmes much more critically. But for governments too there were strong pressures in the same direction. The public embarrassment of huge cost and time overruns on big weapons projects began to be noticed by legislatures and by public opinion already in the 1950s. Moreover, the very fact of the huge expansion carried with it the demand for greater public accountability. As long as R & D budgets were minute they could escape without much scrutiny, but once they were measured in millions and billions, a higher degree of management attention was almost inevitable.

Consequently, during the 1960s two new tendencies became apparent in most OECD countries: first a slowing down in the rate of growth of R & D budgets and secondly far greater interest in the results of R & D. At the same time, academic researchers as well as industrial managers began to take a serious interest in the whole process of industrial innovation and technical change, as well as in the narrower problems of R & D project evaluation and programming.

The first effects of this recognition in the 1960s tended to be a reinforcement of the 'supply-side' policies of the 1950s—a rather indiscriminate build-up of R & D facilities and of the education system. In this early period the 'technological' and productivity gaps between Europe and the United States were the main focus of concern—the gap with Japan only opened up in the 1970s and 1980s, although it was foreshadowed in a few industries, such as ship-building, much earlier. The US lead in the 1950s was often wrongly attributed primarily to the scale of military, space and nuclear R & D and to the government policies affecting these technologies (e.g. Servan-Schreiber, 1965).

However, during the 1960s with the increasing sophistication of economic and political analysis, the 'spin-off' arguments of the military and big science lobbies became increasingly discredited. The US–European productivity gap was steadily closing and studies like that of Eads and Nelson (1971) demonstrated, both in terms of economic theory and of practical examples, the wastefulness of much 'big technology' government-financed development. A series of studies, notably by Carter and Williams (1957, 1959a and b and 1964) in the UK and by Hollander (1965) in the US, pointed to the great importance of other scientific and technical activities, as well as R & D in the industrial innovation process. Thus, although projects such as 'Concorde' and fast breeder reactors still received lavish financial and political support throughout the 1960s and indeed throughout the 1970s, they were subjected to more and more searching examination and criticism and even the expenditure of NASA after the triumph of the moon landings was drastically reduced. Scarcely a single economist could be found in Britain or the United States to lend support to government subsidies for the SST. In part, of course, these more stringent financial approaches simply reflected a general tendency to curtail the growth of government expenditures, which was gathering force in the 1960s and even more in the 1970s. However, this was by no means the only factor as some other types of expenditure on science and technology were increased even during the 1970s. There was a growing sophistication in policies for science and technology associated with a better theoretical understanding of its role in economic growth and international competition (Pavitt and Walker, 1976).

This greater sophistication found expression in many ways—in the development of new institutions for science and technology policy in the 1960s and 1970s in most OECD countries, which were no longer just 'big science' or nuclear organizations, in the debates within the OECD and the ministerial meetings; in projects such as the 'Six Country' project,[2] and in the growth of relevant research and

[2] 'Six-Country Project', an inter-governmental international project sponsored by Canada, France, Germany (FR), Netherlands, Ireland and UK. Many seminars and publications available

the dissemination of the results. Characteristic of all these new developments was a growing concern with cost effectiveness and with the direct contribution of science and technology to enhanced economic performance. The emphasis was far less on the health of the ST system or the RD system and much more on the economic and social environment necessary to make good use of the output of science and technology. Typical questions which preoccupied the new ministers, councils, committees and other institutions were the relative advantages of various methods of funding R & D, and innovation more generally, methods of project evaluation and 'technology assessment', government procurement, the role of the tax system, improvements in communication and information systems, monopoly and competition in relation to innovation policy and so forth. The debate in Britain was particularly intense because of the relatively poor performance of the British economy over a long period (Pavitt (ed.), 1980).

The concentration on the economic growth objective did not go completely unchallenged even though it predominated in most countries in the 1960s (OECD, 1971). It was questioned on a variety of grounds but especially on two—whether it could be sustained in the long term because of the exhaustion of materials and energy resources, and whether the pollution associated with it might not endanger the very existence of the human race. Neither of these questions was completely new but they attracted great attention in the early 1970s and this had a considerable influence on policies for science and technology throughout the OECD area. They were brought into focus by the debate on *The Limits to Growth* (Meadows, 1972) and the counter-arguments advanced in such critiques as *Thinking about the Future* (Cole *et al.* (ed.), 1973). One of the outcomes of the debate was an acceptance by most governments and by public opinion that environmental policies should have much greater weight in decision-making, and that considerable scientific and technical efforts should be devoted to the attainment of higher environmental standards and the prevention of some of the more serious pollution hazards. Although there has been some back-sliding more recently, these remain important influences on the allocation of resources for science and technology and on policy generally, and as we have seen (Table 9.2), did result in a significant reallocation of resources in the 1970s.

Another outcome of the debate was a recognition that if long-term economic growth were to be sustained over a long period, then it could only be through a high rate of technical change in the use of materials, energy and the stock of capital. Critics of the MIT work demonstrated that new discoveries, the use of lower grade ores, economy in use of materials, substitution processes, recycling and social changes could combine to avert the catastrophic scenarios of collapse of the world economy in the next fifty years, but all of this would be possible only if technology policies were successful and integrated with economic policies, and responded fairly rapidly to new developments.

This particular lesson was brought home forcefully by the OPEC crisis of 1973 which confronted the OECD countries with the need to develop urgently alternative sources of energy and at the same time to adopt much more efficient policies for conservation. Both of these requirements also had major implications for priorities in R & D and the introduction of completely new programmes to cope with the situation, such as work on solar energy, geothermal energy, biomass

from TNO, Netherlands. See particularly R. Rothwell and W. Zegveld, *Industrial Innovation and Public Policy: Preparing for the 1980s and 1990s*, Frances Pinter, 1981.

and so forth. Thus the second phase of post-war policies for science and technology in the OECD area ended with a growing recognition that the old objectives of sustained economic growth would be difficult to maintain in the 1980s without some re-orientation of science and technology and without a far closer integration of STP with economic, industrial and energy policies. The problem of technical change and employment came increasingly to the forefront and this is dealt with in the final chapter of this book.

The second stage of STP in the OECD countries featured a less naïve approach to the problems of R & D and innovation. It was marked by the establishment of separate ministries for science policy or technology policy or both in a number of countries or by considerably strengthened groups at Cabinet level in others. At the same time statistics of R & D became generally available and were put on an internationally comparable basis, largely through the efforts of the OECD. This organization also initiated a series of national science policy 'Reviews' in each of the member countries, which heightened awareness and understanding of the problems. Finally, within the academic world increasing attention was devoted to the economic and social aspects of science and technology. Some of the leading technological institutions in the world, such as MIT and Imperial College, began to develop significant research and teaching programmes in this area, while the Science Policy Research Unit at Sussex and similar groups in several of the OECD countries had already shown that it was possible for academic research to make a useful contribution to both national and international policy debates.

The achievements of the second stage of STP should not be overrated, any more than those of the first stage. Although policy-makers for science and technology became aware of the high opportunity cost of military R & D and some other types of big science and technology, they were unable or unwilling to achieve very much in the way of re-orientation of these expenditures. During the 1960s, military expenditures were falling in most countries as a proportion of total R & D expenditures, but in the 1970s they began to creep up again. The Concorde project was never actually cancelled in the 1970s, despite the overwhelming evidence that it could never be economically viable, and there were other smaller 'Concordes' elsewhere. Even where resources were transferred or new resources were invested in programmes which were likely to have a more direct and beneficial effect on economic efficiency and economic growth, there were great difficulties in organizing and administering such programmes and considerable doubts about their value. Although new governmental structures were set up in many countries to deal with STP problems, quite often they did not endure for very long and there was a bewildering succession of new committees, councils, agencies, ministries etc. Although social studies of science and technology gained in academic respectability (Kennedy and Thirlwall, 1971) and textbooks of economics began at last to include chapters on technical change and R & D, still the mainstream of teaching and research contrived to ignore these issues most of the time.

The reasons for these failings go deep to the heart of the economic, social and cultural structures and traditions of most OECD countries. They concern the relationships between government and industry in a market economy, between the various functions of government in departmental systems of administration, and between the various disciplines in an educational and career system organized on rigid disciplinary lines. They also involve the attitude of the labour force and management to technical change and the responsibility for introducing and managing it.

The third stage of post-war science and technology policy

Science and technology policy cannot solve these fundamental social problems. Few people today believe in the notion of the scientist-rulers or technocracy, which was common in the nineteenth and early twentieth century and was popularized by writers like H. G. Wells (Mackenzie, 1973) and scientists like J. D. Bernal (1929) in his youthful masterpiece *The World, the Flesh and the Devil*. Bernal himself explained in his own later work (1939) why this notion was no longer acceptable or realistic for most people: first, there was no evidence that the population in general would be ready to entrust scientists with this responsibility, and secondly, there was no evidence that scientists would be able to make a better job of it than anybody else—it would much more likely lead to some form of fascism.

Nevertheless, the idea that somehow 'science' ought to be able to provide all or most of the answers, and that the social sciences could somehow or other transcend the clash of interest groups and the passions of ideological conflicts keeps re-surfacing in one form or another. Most recently, it has often done so in the form of a new technique such as systems analysis or computer modelling, or in an appeal to 'Popperian' logic and scientific method or positivism in the social sciences. Particularly in Eastern Europe, systems analysis is sometimes seen as a substitute for political debate and conflict.

However, despite these recurring tendencies, most people are reconciled to the view that the political system is and will remain the arena of conflicting interests and opinions and considerable uncertainty about the way in which human affairs ought to be organized. If this is the case, then 'science policy' cannot aspire to a quality of universality any more than any other policy-making. It can, however, certainly take into account the interdependence of economics, social affairs, ethics, foreign policy and technology and science. The fact that we do not have a universal prescription or diagnosis for all our social ills does not mean that everyone should retreat into a fragmented view of the world, whether this be based on a narrow individualism, on departmental attitudes, or the blinkered view of a single discipline or paradigm, or a naïve belief in the perfection of the market.

Somehow or other, we have to make the best sense we can of a disordered and imperfect world, and like any good research project, grope towards solutions of which we are only half aware.

The science and technology system has shown its capacity to respond fairly rapidly to a great variety of new demands. This applies both to private industry and to the public sector, as well as to the universities. Certainly there are problems of mobility, adaptability and substitution of various categories of scientific and technical manpower. But the 'big science' and 'big technology' programmes of the 1950s and 1960s have at least demonstrated unequivocally that there is far more flexibility and capacity for swift response to changing demands than had often been assumed and this was again shown by the response to the energy crisis of the 1970s. Science and technology are already contributing to the solution of many environmental, development and welfare problems, and they are certainly capable of meeting much greater demands. The problem for policy is to articulate these demands in such a way that the system can respond effectively.

Strategic priorities

The direction of research priorities today will in large measure determine the range of real choice available to consumers in future decades. Consequently this is a question of fundamental importance in any democracy and the Rothschild approach of leaving the main priorities entirely to departmental decision-making is unacceptable. Even less acceptable is the primitive 'know-nothing' mentality in some branches of the government statistical services which attempted to suppress the collection and publication of R & D statistics, and the perverse secrecy of much civil service discussion on new priorities in research and technology.

To look at expenditure on scientific research and experimental development on a global basis does make sense, even though decision-making is largely decentralized and sometimes industrial R & D is quite independent of government. This does not necessarily imply that there is or ought to be a 'science budget'—a question which has been a matter of inconclusive debate for some time (see, for example, OECD, 1966 and UN Economic Commission for Europe, 1968). It *does* imply that some guidelines are needed for the main priorities in public expenditures for science and technology and for industrial innovation. The question of the scale of expenditure on space research, medical research, military and civil aircraft, particle accelerators, ocean research and so forth cannot be decided entirely on a departmental basis, looking at each case in isolation. Since resources are limited, opportunity costs must always be taken into account. Even if they were ill-considered, there were social priorities clearly implicit in the pattern of R & D expenditures in the 1950s, 1960s and 1970s. This applies both to the public and the private sectors.

The advance of science and technology must find its support and its justification, not merely in the expectation of competitive advantage, whether national or private, military or civil, but far more in its contribution to social welfare, conceived in a wider sense. The funding of R & D is extremely important for these basic goals and the *strategic* aims of research and innovation, i.e. *policy* for research, may often be more important than particular projects. This is clear from chapters 7 and 8 as well as from military and space experience.

To modify the flow of funds to research and development in such a way that they contribute more directly to the goals of social welfare and environmental improvement may not prove so difficult. It is, after all, only to reinforce trends which are already apparent in the industrialized countries. Far more difficult will be the development of institutions to assess, modify and direct technical progress in such a way as to realize the full benefits of this research, and to ensure that the social innovation mechanism functions effectively. 'Technology assessment' represents the greatest challenge both to the political system and to the social sciences in measuring, representing, displaying and imagining the benefits and costs of new technologies.

Public participation in the process of consumer-oriented innovation has very great implications for the education system as well as for the political system, the mass media and science journalism. Television programmes, such as 'Tomorrow's World', are already playing an important part, but they will require to be complemented by other methods, such as that used in the 'Orakel' system of audience participation in the evaluation procedures for alternative technological possibilities (Krauch, 1972). The function of 'technology critic' is just as important as the function of 'literary critic' or 'art critic' and to some extent these should overlap.

But such criticism will be far more effective when social scientists develop quite new techniques of technology assessment.

Present R & D project selection techniques are biased overwhelmingly towards technical and short-term competitive economic criteria. This is true of both capitalist and socialist economies. An extremely important problem for research and social implementation is the evolution of quite new techniques of selection and assessment which could be applied both in the private and public sectors. These should take into account aesthetic criteria, work satisfaction criteria, environmental criteria and other social costs and benefits which today are almost excluded from consideration.

One of the most harmful consequences of the division of labour has been the absurd attempt to set up the 'arts' and 'sciences' in opposition to one another. They are mutually complementary modes of apprehending and changing the universe and society, which need each other. Just as an individual combines intellect, imagination, reason, feeling and intuition, so must a healthy society blend the fragmented compartments and artificial divisions in our knowledge system and our professions. Only in this way can alienated technology become human technology. Innovation is far too important to be left to scientists and technologists. It is also far too important to be left to economists or social scientists. This leads to a discussion of the general problem of 'consumer sovereignty' in relation to innovation, and the ability of both the political system and the economy to assess the needs and desires of the population.

Innovation in consumer goods and services

The achievements of the military innovation system are undoubted. The evidence of Part One has also shown the very great technical achievements of innovators in capital goods, components and materials in the twentieth century. The *buyers* of capital goods and intermediate products, as of weapon systems, are often scientifically and technically sophisticated customers. They are their own 'customers' for many process innovations, and when they are purchasing outside they often have greater technical sophistication than their suppliers (for example the chemical firm selecting a pump or a filter). They may often use their purchasing power to commission technical innovations. Buyers in these markets are concerned with genuine technical characteristics and may lay down stringent technical performance specifications. They are less likely to be impressed by product differentiation, and advertising plays a less important part than with consumer goods while technical services to customers play a much more important part.

Most innovation case studies agree that those innovators who take considerable trouble to ascertain the future requirements of their customers are on the whole more successful. The SAPPHO comparisons of success and failure in innovation showed that most failures were associated with either neglect of market requirements or relatively poor understanding of the customer's needs. To this extent the argument here is at variance with Galbraith's assumption of 'producer sovereignty' in imposing innovation on the market.

These results are superficially encouraging in so far as they indicate that at least in capital goods the market has been effective in stimulating the types of innovation which match real customer needs and—potentially, if not always in practice—social welfare. There are certainly also examples of *consumer* goods which confirm the SAPPHO conclusions on user needs. Take, for example, the

Danish plastic toy, 'Lego'. This firm took an enormous amount of trouble to ascertain the needs and preferences of the users (in this case, mainly children). It has been rewarded by the most successful sales and export performance of any toy in the world. On the only occasion when the firm introduced a new product without its usual exhaustive prototype tests with users, the innovation was a failure. There are many other examples of great consumer benefits from household products and from drugs. Moreover, most consumers have benefited from the rise in living standards made possible by the productivity advances due to technical innovation in capital goods, materials and communication systems.

Nevertheless, generalizations about the benefits of technical innovation do need to be heavily qualified outside the area of capital goods. Most innovation studies have been concerned with the more spectacular 'break-through' innovations and have hardly considered the type of 'annual model' changes which are more characteristic of many consumer products. There are reasons for believing that buyers in these areas are far less capable of making sound technical judgements than in the capital goods area. They may typically have rather poor sources of information and lack the capacity to make any serious technical assessment.

A glance at the distribution of industrial R & D expenditures in any country shows a heavy concentration in capital goods and materials. Consumer industries have very little and much of these limited R & D and other scientific inputs are used for product differentiation in oligopolistic markets, and in the closely related activity of planned obsolescence. The growth and welfare implications of this kind of industrial R & D are very dubious. Fisher, Griliches and Kaysen (1962) have shown the very high costs of annual model changes in the US car industry and the 'planned obsolescence' in car exhaust pipes was strongly criticized in a UK government report (Report of the Committee on Corrosion and Protection, 1971, p. 128). Obviously it was this aspect of technical innovation which Galbraith had in mind in developing his critique of 'producer sovereignty'. This critique is much more relevant to consumer goods than to capital goods.

This raises the more general problem of the economic theory of the market and the direction of innovation. Theoretically, the ideal consumer market is supposed to provide consumers with the power to choose between an array of alternatives. Possessing 'perfect' information they are free to choose the 'best buy' for price and quality, thus compelling suppliers to adapt their output to consumer needs through the competitive mechanism. This was the idea of 'consumer sovereignty'. Of course, no economist ever imagined that reality would ever quite correspond to the ideal. Consumers would never really be 'perfectly' informed and very frequently there would be some form of collusion between suppliers or other elements of monopoly. But in some primary commodity markets, fruit and vegetable markets, and some markets for capital goods the reality has sometimes approximated quite closely to the ideal abstract model.

Economists have devoted a great deal of attention to the problems of maintaining consumer sovereignty (see, for example, Knox, 1969). Much of the theory of monopoly and the critique of advertising has been concerned with the erosion of consumer choice. In many countries anti-monopoly and consumer protection legislation attempts to restrict or reverse the powerful tendencies towards concentration of producers and greater 'producer sovereignty' (Heath, 1971). We are not concerned here so much with this *general* problem but with the *specific* consequences of *technical* innovation on consumer markets.

There are three main ways in which technical innovation may diminish 'consumer-sovereignty' and which the normal type of anti-monopoly legislation does little to affect:

1. The theory of consumer choice is essentially *static*. The consumer supposedly chooses from the *existing* array of goods and services. But in areas where technical change is important this array has been determined by choices of R & D project or innovation decisions many years before. The critical element lacking in 'consumer sovereignty' is therefore the power to influence the *future* array of goods and services. Apologists for the present state of affairs maintain that consumers do in fact exert this influence *indirectly* because producers are concerned to anticipate their wishes in order to make a profit. Up to a point this is true, as we have seen in the case of many products, but the possibility exists that the innovation decision-makers will impose *their* preferences and *their* choices, rather than those of the consumer. This power is not unlimited, as the example of the Ford Edsel showed, but it is a serious problem.

2. The theory of consumer choice implies perfect information about the available array of products or services. If we are thinking of a housewife looking at the prices and quality of vegetables on a dozen different stalls in a street market this model may not be too far from the truth. But it breaks down where any degree of technical sophistication or product differentiation come into the picture, as with cars, television and other consumer durables, and increasingly with a great variety of processed food and chemical products. This market situation is essentially one of completely unequal access to technical information, as all the consumer associations are well aware. The extremely unsatisfactory nature of the repair and maintenance services for consumer durables in all industrial countries is one illustration of this problem. Despite the fact that there are thousands of garages competing for the motorist's custom, consumer dissatisfaction is chronic and well-founded.

3. Consumers cannot possibly be aware of the various long-term side-effects of a multitude of individual choices about many products. The individual who buys a car did not and could not work out the long-term consequences for the urban environment of millions of similar decisions. Nor indeed did the suppliers. Yet the social costs may well be so great that they negate the private benefit to most consumers, as Mishan has so persuasively argued (1969). The problems of waste disposal in relation to plastics and nuclear power are other examples of long-term social costs inadequately considered.

These three defects help to explain why the nominally sovereign consumers may often feel powerless and frustrated. The problem is often one of adequate articulation of a felt need. But it may often also be that producers are almost exclusively oriented to product differentiation and brand image, rather than to imaginative technical innovation or to social needs. For example, efforts to design safer cars and a relatively pollution-free car were very obvious social needs, but the R & D effort which went into improving safety or preventing pollution was negligible for a long time. Interestingly enough, the stimulus came not from the R & D of producers but from outside critics, such as Ralph Nader, and public regulation. The same is even more true of designing a generally more satisfactory land transport system, particularly in congested urban areas.

Several other examples may be cited to illustrate the extent to which innovators and designers may neglect the interests of users and simply pursue their own fashions and enthusiasms. The example of housing is particularly striking in Britain.

In the 1960s, there was a fashion among architects and town-planners for high-rise flats ('high' blocks of flats may be defined for this purpose as blocks with six or more stories). The extraordinary feature of this particular technological fashion was that such flats were much more expensive than conventional housing or low-rise flats throughout this period (McCutcheon, 1972). (Estimates vary between 1.3 and 1.8 times the cost per square foot.) Moreover, this was a period of very great financial stringency when local authorities were under constant pressure from central government (and from electors) to prune their expenditures and cut back on their services. Yet in spite of this almost all major cities put up thousands of high flats throughout the 1960s.

This might have been justifiable if there had been overwhelming evidence of consumer preference for high-rise living, from sociological surveys and/or a clear readiness to pay the much higher rents on an economic basis. No such evidence was ever produced. The sociological surveys were inconclusive but most of them showed, if anything, a dislike of high flats among council tenants, particularly those with children and old people.

These examples demonstrate the extent to which the values and preferences of designers and innovators may be *imposed* on the consumer, whether through private firms or public authorities. This does not imply malevolence or contempt for the consumer. On the contrary, in every case, the innovators believed that they were acting in the best interests of the consumer. It is only a special illustration of a problem which has long been familiar to political scientists. The separation of research, development and design into specialized functions with their own ethos, fashions, interests and enthusiasms, inevitably carried with it this danger of lack of social accountability. In theory again the competitive market mechanism ought to be able to perform this function automatically. But it has been argued that the market place mechanism, which theoretically was supposed to ensure correspondence between consumer wishes and supply, no longer performs this function adequately, if it ever did so in some sectors. This means that increasingly the *political* mechanism must restore the lost consumer sovereignty which the autonomous market mechanism can no longer assure.

It might have been expected that socialist societies would have been able to make social innovations, which would link the public R & D system more closely to consumer needs. But the evidence available does not justify this conclusion, possibly because they have been poor countries attempting to industrialize rapidly, and in the case of the Soviet Union and China, to compete militarily with other great powers.

There seems no reason in principle why the more enterprising consumer associations should not go into the business of specifying desired technical performance parameters in the same way as capital-goods buyers. They might do this either in association with public authorities or on their own, depending upon circumstances. Ultimately this could even lead to the award of R & D contracts and procurement contracts. Some chain stores and department stores already act in this way on behalf of their customers and, in closing this loop, the civil sector would only be learning the social technique which was learnt by the military sector a generation ago, and by capital goods innovators long ago.

The 'customer–contractor' principle of the Rothschild Report (1971) can be of great value if it is interpreted in this sense—as an assertion of the need for all R & D organizations to operate with a sense of social responsibility, and a way of improving the 'coupling mechanism' between the scientific community and the public. However, the Report tended to assume a one-to-one correspondence between the interests and preferences of government departments and the interests and preferences of the public on whose behalf they are acting. This is of course very far from being the case. Departments may often share the technological fashions and preferences of designers without checking back to the actual wishes of consumers. (This was in fact the situation both in the case of aircraft development and of high-rise flats.)

One must certainly hope that the executive branch will be responsive to the known wishes and needs of the population, and will be active in trying to ascertain these preferences where they are not known. But the experience of politics over several thousand years has shown conclusively that this cannot be relied upon (Armytage, 1965; Lenin, 1917). Methods are therefore essential which can ensure that the executive branch is subjected to a continuous process of critical review and control. Local and spontaneous initiatives are important but so too is Parliament, and the strengthening of Parliamentary process in relation to science and technology is essential.

This final chapter first of all returns to a synthesis of some of the conclusions emerging from the first part of this book. These pointed to the crucial role of the emergence of new industries and technologies in the long-term growth of the economy and to the increasing role of industrial R & D in that process. At the same time, they emphasized the extremely uneven distribution of R & D activities between different branches of industry and the stages of accelerated growth which were associated with the diffusion of major innovations.

This discussion was not related to business cycle phenomena nor to the issue of unemployment, but this chapter briefly explores some of these connections. Since the issues are complex, before taking up the specific problems of technology *policy*, it is first necessary to clarify some of the basic theoretical issues, paying particular attention to the ideas of Schumpeter, who in this field too made an outstanding original contribution.

Unemployment in the early 1980s in most of the industrialized countries was higher than at any time since the 1930s. This was not merely the result of short-term fluctuations, but reflected a secular trend associated with generally slower rates of growth, slackening investment and more determined efforts to contain the growth of government expenditure. The period of very rapid economic growth which succeeded World War II gave way to a rather different phase of recession and 'stagflation'. In these circumstances it is hardly surprising that interest revived in theories of Kondratiev 'long cycles' or 'long waves' in economic development, which seek to explain such long-term changes in the economic climate. Joseph Schumpeter, more than any other twentieth-century economist, attempted to explain these cycles in economic growth largely in terms of technical innovation.

Schumpeter's theory of long cycles

There *do* appear to have been in the past century or so several periods of rather deep crises and slower growth, associated with higher levels of unemployment, which at least in some countries were regarded at the time, and by historians since, as 'Great Depressions'. These alternated with the periods of boom and prosperity experienced in the 1850s and 1860s, in the *Belle Époque* before World War I and in the 1950s and 1960s. The depressed periods were roughly the 1880s and the 1930s, although varying a little from country to country. In the United States, for example, the Civil War in the 1860s was followed by a period of rather rapid growth in the 1870s and 1880s, when European countries were generally experiencing slower growth rates and more depressed conditions. Japan was less affected by the depressions of the 1930s and (so far) the 1980s.

However, to appreciate the relevance and importance of Schumpeter's ideas on technical innovations and economic fluctuations, it is by no means necessary to accept the idea of cycles as such and certainly not the notion of fixed periodicity. Van der Zwan (1979) prefers the notion of periodic major structural crises of adjustment and Mensch (1975) speaks of a 'metamorphosis' model. Almost all the advocates of the idea now prefer to speak of 'long waves' rather than 'long cycles'. For the purpose of this book it is necessary only to accept van der Zwan's concept that there have been periodic major structural crises of adjustment varying a little in their severity and timing between countries and followed by fairly prolonged periods of expansion and prosperity. These major crises were generally perceived at the time as being more severe than the ordinary down-turns of the short- and medium-term (Jugler) business cycles.

In his major work on *Business Cycles*, Schumpeter (1939) both accepted the reality of the phenomenon of 'Kondratiev'[1] long cycles, lasting half a century or so, and offered a novel explanation of them, differing from that of Kondratiev (1925) himself. According to Schumpeter (1939, Chapter 2), each business cycle was unique because of the variety of technical innovations as well as the variety of exogenous events such as wars, gold discoveries or harvest failures. But despite his insistence on the specific features of each fluctuation and perturbation, he believed that the task of economic theory was to go beyond a mere catalogue of accidental events, and analyse those features of the system's behaviour which could generate fluctuations irrespective of their specific and variable form. The most important of such features in his view was innovation, which, despite its great specific variety, he saw as the main engine of capitalist growth and the source of entrepreneurial profit.

The ability and initiative of entrepreneurs (who might or might not themselves be inventors but more usually would not be) created new opportunities for profits, which in turn attracted a 'swarm' of imitators and improvers to exploit the new opening with a wave of new investment, generating boom conditions. The competitive processes set in motion by this 'swarming' then gradually eroded the margins of innovative profits (as in Marx's model), but before the system could settle into an equilibrium condition, the whole process would start again through the destabilizing effects of a new wave of innovations. This process was sufficient in itself to engender various types of cyclical behaviour, although Schumpeter certainly acknowledged that there was a process of interaction with many other features of the economic system which have been the subject of more conventional business cycle analysis.

Whether or not such a mechanism offers a plausible explanation of 'long' (Kondratiev) cycles in economic development depends crucially—as Kuznets (1940) pointed out in his review of *Business Cycles* at the time—on whether some innovations are so large and so discontinuous in their impact as to cause prolonged perturbations or whether they are bunched together in some way. The construction of a national railway network might be the type of innovation investment which would qualify as a 'wave generator' in its own right, but obviously there are thousands of minor inventions and technical changes which are occurring every year in many industries whose effect is far more gradual and which might well adapt to some sort of smooth equilibrium growth path. If these smaller innovations were to be associated with economic fluctuations, then this could only be if they were linked to the growth cycles of new industries and technologies.

A neo-Schumpeterian interpretation of the post-war boom would see it primarily as the simultaneous explosive growth of several major new technologies and

---

[1] As has often been pointed out, Kondratiev was by no means the originator of the long cycle theory and it is in some respects a misnomer that the phenomenon bears his name. The Dutch Marxist van Gelderen (1913) could be much more fairly credited with the idea, which he articulated clearly in 1913. At about the same time a variety of economists, including Pareto (1913), had drawn attention to the apparent tendency for long-term price movements, interest rates and trade fluctuations to follow a cyclical movement lasting about half a century. However, during the 1920s whilst heading the Institute of Economic Research in Moscow, Kondratiev did more to propagate and elaborate the idea than any other economist.

industries, which have been discussed in Part One, particularly electronics, synthetic materials, drugs, oil and petro-chemicals, and (especially in Europe and Japan) consumer durables and vehicles. It can be shown that these fast-growing industries and the associated component and machinery suppliers accounted for a large part of the growth of industry in many OECD countries. It is interesting to note that neither Keynes nor most Keynesians expected that the quarter century after the war would be the fastest period of economic growth the world had ever experienced. Indeed the expectations of economists immediately after World War II were mostly pessimistic. This was partly because they tended simply to extrapolate the problems of the 1920s and 1930s into the post-war world and made no allowance for new technologies and the impetus they would give to profit expectations and to new investment. When, however, the big boom did materialize, the tendency was to attribute its success to the adoption of Keynesian policies. An important exception was Robin Matthews (1968), who already in 1968, in an article in the *Economic Journal* on why Britain had had full employment since the war, pointed out that it was the buoyancy of private investment rather than government policies which sustained the boom. Other economists (Dow, 1964) pointed out that Keynesian policies were often inept since they were applied after the relevant turning-point in the business cycle had already occurred 'spontaneously'.

Schumpeter had suggested that after a strong 'band-wagon' effect and the entry of many new firms into the rapidly expanding sectors attracted by the exceptionally high profits of innovation, there would follow a period of 'competing away' of profits as the new industries matured. This could lead to stagnation and depression if a new wave of innovations did not compensate. Such an explanation appears to fit the facts of the post-war boom. It had been remarked already in the 1960s that the general rate of profit was beginning to fall in several OECD countries and this tendency was aggravated by more severe international competition. This became still more marked in the 1970s, especially in some of the erstwhile rapid growth sectors such as synthetic materials and consumer durables. In spite of its continued high rate of growth, the electronics sector was also affected by this general trend.

Schumpeter had relatively little to say about unemployment and wages, but economists of many different schools, including both neo-classical and Marxist, would agree that a sustained period of boom and full employment such as that experienced before World War I or after World War II would tend to strengthen the bargaining position of labour and thereby also to erode the rate of profit and stimulate cost-push inflation, with or without the pressure of militant trade unions. These changes would tend to induce rather different types of technical change, and associated investment. In the early period of a long boom the emphasis is on rapid expansion of new capacity in order to get a good market share, and this investment has a strong positive effect on the generation of new employment. As the new industries and technologies mature, they are standardized, economies of scale are exploited and the pressures shift to cost-saving innovations in process technologies. Capital-intensity increases and employment growth slows down or even stops altogether. Again, this hypothesis appears to fit the post-war pattern quite well. Already before the OPEC crisis these trends were apparent in the leading industrial countries, both in the newer industries such as synthetics and electronics, and in older ones.

In Schumpeter's framework it is disequilibrium, dynamic competition (in the sense of 'imperfect' competition) among entrepreneurs, primarily in terms of industrial innovation, which forms the basis of economic development. Thus, the emphasis is on the supply side, that is, autonomous investments rather than on 'demand induced accelerator investments or multiplier processes [demand push] as driving forces in economic development' (Giersch, 1979, p. 630). In such a framework economic development will be viewed primarily as a process of reallocation of resources *between industries*. That process leads automatically to *structural* changes and disequilibria, if only because of the uneven rate of technical change between industries. In other words, economic growth is not only accompanied by the rapid expansion of new industries; it also primarily depends on that expansion.

In his insistence on the importance of the specific features of each perturbation to the system, Schumpeter differed from most Keynesian theorizing and from much other macro-economic crisis theory with its blithe disregard of actual technical developments and real changes in the structure of industry and services. Like Marx and Kuznets, his approach was evolutionary and historical and he was constantly aware of the limitations and dangers of abstract generalizations and models. He saw research on 'business cycles' as having to rely on company histories, technical journals, studies of particular products and branches of industry, and repeatedly emphasized how misleading aggregative statistics could be, since they frequently concealed rather than revealed the underlying processes of change.

He justified his view that technical innovation was 'more like a series of explosions than a gentle though incessant transformation' on three grounds. First, he argued that innovations are 'not at any time distributed over the whole economic system at random, but tend to concentrate in certain sectors and their surroundings', and that consequently they are 'lop-sided, discontinuous, disharmonious by nature'. Secondly, he argued that the diffusion process was inherently a very uneven one because 'innovations do not remain isolated events, and are not evenly distributed in time . . . on the contrary, they tend to cluster, to come about in bunches, simply because first some and then most firms follow in the wake of successful innovation.' Thirdly, he maintained that these two characteristics of the innovative process implied that the disturbances it engendered could be enough to 'disrupt the existing system and enforce a distinct process of adaptation' (see Fels, 1964, pp. 75–7).

Hardly anyone would deny the truth of Schumpeter's first proposition. As we have seen in Part One, it is confirmed by a great deal of empirical research on the uneven distribution of R & D, patents, inventions and innovations between the various branches of the economy. The differences between rates of growth of various branches of production are well known and obvious, as is the fact that some industries decline whilst others grow rapidly.

The most R & D intensive industries are those with extraordinarily high rates of growth in the twentieth century and most of them did not exist at all before this century—electronics, aircraft, drugs, scientific instruments, synthetic materials (chapter 1). Again, it is fairly obvious that these high rates were related to a much greater flow of technical innovation in new products and processes and the high rate of diffusion of these innovations through the world economy. Differences in growth rates of production and of productivity have been systematically related to R & D intensity and to patterns of technical change (Terleckyj, 1974).

At the other extreme are industries in a process of decline or showing rather low rates of growth. These are frequently characterized by low R & D intensities (sometimes zero) and a low rate of technical change. Where technical innovation is taking place, as, for example, in the textile and printing industries, it is often due more to the impetus from suppliers of machinery and materials, than from the efforts of the industry itself. The existence of a statistical association between measures of technical change and the growth of an industry or product group does not, of course, necessarily mean that it is technical innovation which has caused the growth. The reverse could be true, or both could be ascribed to some other factor, such as the quality of entrepreneurship or market demand. Schumpeter stressed the importance of autonomous invention and entrepreneurship, but Schmookler (1966) put the main emphasis on market demand.

Schumpeter or Schmookler?

However, if market demand were the only problem, then technical innovation could be regarded as a secondary phenomenon and taken for granted, since it would simply respond to demand management. It would be part of the adjustment to changing patterns of demand, not a semi-autonomous engine of growth. Schmookler's theory has sometimes been interpreted in this way. Fluctuations in investment are followed by fluctuations in inventions. In Part One some evidence has been presented which suggests that a purely demand-led theory of invention and innovation does not correspond to the historical facts in the case of the technologies which are discussed. Schumpeter's theory of an autonomous impetus on the supply side, deriving from advances in science and invention and realized through imaginative entrepreneurship, appears to fit the facts rather better. As we have seen, however, once a major innovation has been made, then a pattern of demand-led secondary inventions and innovations may set in over many decades giving apparent credibility to a 'Schmookler'-type of analysis. Granstrand (1982) has pointed out that the reverse process sometimes occurs when a renewed technology push leads to new applications and demands.

Dosi (1982) has suggested an interesting parallel between technological paradigms and Kuhn's theory of scientific paradigms. A major new technology is comparable to a new paradigm in science; as the technology takes off, its further articulation is the outcome of 'natural trajectory' possibilities (comparable to normal science in Kuhn's theory), heavily influenced by the market selection environment. Technical change is both an engine and a thermostat, but the thermostat function tends to predominate as a technology matures.

Chapter 3 showed that the evidence from the plastics industry, whether descriptive or statistical, does not support either a deterministic model of demand-led invention or of technology-led science. Neither would it sustain a theory of pure 'discovery push' or invention push, which ignored the reciprocal influence of the growth of demand, of fluctuations in economic activity and of competitive pressures. But it would be consistent with a theory such as Schumpeter's which postulated a closely interdependent but shifting relationship changing over time as an industry grew to maturity. Exogenous science and new technology tend to dominate in the early stages, whilst demand tends to take over as the industry becomes established. A 'matching' process of new technology and new markets, guided by imaginative entrepreneurs, is important throughout.

However, as we have seen, Almarin Phillips (1971) pointed out that there is not one Schumpeterian model but two. The first is that already developed by the young Schumpeter before World War I and expounded in his *Theory of Economic Development* (1912). The second is that advanced in his later book *Capitalism, Socialism and Democracy* (1942). Figures 10.1 and 10.2 are a schematic

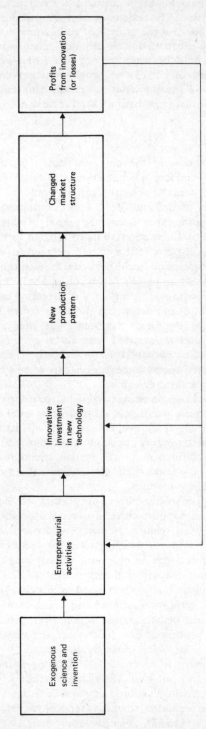

Fig. 10.1   Schematic representation of Schumpeter's model of entrepreneurial innovation (Mark I).

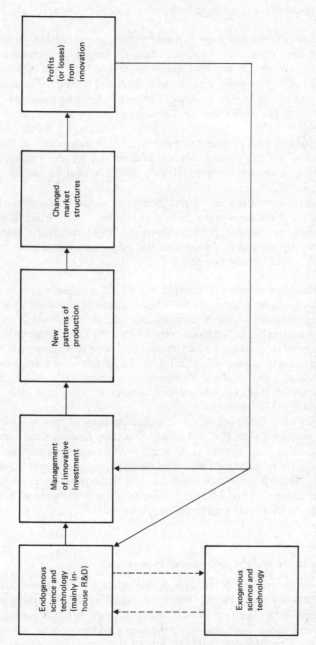

Fig.10.2   Schematic representation of Schumpeter's model of large-firm managed innovation (Mark II).

representation of these two models which have been designated as Schumpeter I and Schumpeter II. They are based essentially on the diagrams used by Phillips (1971) with minor modifications. The pattern postulated by the Schumpeter I model and illustrated in Fig. 10.1 may be summarized as follows:

(i) There is a (discontinuous) flow of basic inventions related in an unspecified way to new developments in science. These are largely exogenous to existing firms and market structures, and hence to any measurable type of 'market demand', although they may certainly be influenced by the belief in a potential demand or concept of unmet need, or shortages of existing products.

(ii) A group of entrepreneurs (who in Schumpeter's view are responsible for the main dynamic thrust in capitalist economies) realize the future potential of these inventions and are prepared to take the risk of developing and innovating. This hazardous activity would not be undertaken by the average capitalist or manager but only by exceptional individuals, whom he defines as entrepreneurs.

(iii) Once a radical innovation had been made, it would disequilibrate existing market structures and reward the successful innovator with exceptional growth and temporary monopoly profits. However, this monopoly will be later whittled away by the entry of swarming secondary innovators giving rise to the cyclical phenomena already described.

The main differences between Schumpeter II and Schumpeter I are in the incorporation of *endogenous* scientific and technical activities conducted by large firms. There is a strong positive feedback loop from successful innovation to increased R & D activities setting up a 'virtuous' self-reinforcing circle leading to renewed impulses to increased market concentration. Schumpeter now sees inventive activities as increasingly under the control of large firms and reinforcing their competitive position. The 'coupling' between science, technology, innovative investment and the market, once loose and subject to long time delays, is now much more intimate and continuous.

A general long-term tendency towards concentration of innovations in larger firms is quite consistent with the possibility that long-term cyclical upswings are associated with a resurgence of 'Model 1' small-firm innovations and, as we saw in chapter 6, the pattern in electronics and scientific instruments does provide evidence of this. Especially in the birth of new industries, small firms may be exceptionally important. There is further evidence of the resurgence of fast-growing small innovation firms in such new technologies as information processing and small computers.

Clusters of innovations and the diffusion of innovations

The diffusion process cannot be viewed as one of simple replication and carbon-copy imitation (Rosenberg, 1976), but frequently involves a string of further innovations—small and large—as an increasing number of firms get involved and begin to learn new technology and strive to gain an edge over their competitors. None the less, it is this diffusion, with or without further innovations and improvements, which alone can give rise to significant economy-wide effects on the pattern of investment and employment. Consequently, it will sometimes be the case that the basic innovations which have a major impact in a particular long upswing will actually first have been made in a different Kondratiev cycle altogether. This will apply *a fortiori* to the international diffusion of technology.

The 'swarms' which matter in terms of their expansionary effects are those associated with a set of interrelated basic innovations, some social and some technical, and concentrated very unevenly in specific sectors. The 'band-wagon effect' is extraordinarily important—it is the main explanation of the upswings in the long waves. It is the steep part of the 'S-curves' characteristic of many diffusion processes, not the relatively flat piece of the curve which often follows the basic innovation for a few years. The band-wagon effect is a vivid metaphor and it relates to a rapid diffusion process which occurs when it becomes evident that the basic innovations can generate super-profits and may destroy older products and processes.

Here, it is important to note an important development in 'diffusion' theory. During the 1960s, a 'standard model' of diffusion of innovations was developed by Mansfield (1961) and others, emphasizing the role of profitability for potential adopters, the scale of investment required for adoption and the learning process within the population of potential adopters as the key determinants of the diffusion or adoption process. This model, although very useful for many purposes, neglected changes in the environment during the process of diffusion and changes in the innovation itself during that process. In the twenty years since the development of this model, a good deal of empirical work on diffusion of innovations has provided a better basis for generalization in this previously neglected area. Several authors, such as Davies (1979) and Stoneman (1976), pointed to the importance of the *supply* side, and Mansfield himself has emphasized this (1977).

In a seminal paper, Metcalfe (1981) stressed the role of profitability for *suppliers* (as opposed to adopters), and the influence of secondary innovations affecting profitability both for suppliers and adopters:

> Gold (1981) and Davies (1979) . . . have argued that observed diffusion paths primarily reflect changes in the innovation and adoption environment rather than a process of learning within a static situation. As Gold observes, the standard diffusion model rests on the implicit static assumption that the diffusion levels reached in later years also represent active adoption prospects during earlier years. . . .

The demonstration of profitability for suppliers is just as important as the demonstration of profitability for adopters, as only this will normally induce the expansion of new capacity and skills necessary to sustain a rapid adoption process. Emphasis on the role of 'change agents' is in this respect more realistic than the predominantly demand-oriented standard economic model.

In Schumpeter's model, the profits realized by innovators are the decisive impulse to surges of growth, acting as a signal to the swarms of imitators. The fact that one or a few innovators have made exceptionally large profits does not mean, of course, that all the imitators will do so. It is enough that they hope to, or even that they hope to make a fraction of them. Nobody else made such profits from nylon as Du Pont or such profits from computers as IBM. Indeed, some would-be imitators made losses. This is an essential part of the Schumpeterian analysis. As the band-wagon begins to roll, profits are competed away and some people fall off the wagon. Schumpeter himself stressed that changing profit expectations during the growth of an industry are a major determinant for the sigmoid pattern of growth. As new capacity is expanded, at some point (varying with the product in question), growth will begin to slow down. Market saturation and the tendency for technical advance to approach limits, as well as the competitive effects of swarming and changing costs of inputs, may all tend to reduce the level of profitability and with it the attractions of further investment. The impact of incremental technical improvements would tend to diminish gradually in accordance with Wolff's Law.[2] Exceptionally, this process of maturation may take only a few years, but more typically it will take several decades and sometimes longer still.

Diffusion, process innovation and social change

So far it has been emphasized that what matters from the standpoint of large-scale economic fluctuations is not so much the date of a particular basic innovation as a constellation of circumstances favourable to the exceptionally rapid growth of one or more new industries, each involving the combination of a number of related inventions, innovations and economic and social changes. Schumpeter also emphasized managerial and organizational innovations. These may often be just as important as the technical changes for the growth of an industry or technology. Thus, for example, many applications of the steam engine required the reorganization of production on a factory basis, which was an extremely painful and difficult social change at the time. Some of the applications of robotics and micro-electronics, both in manufacturing and service industries, may similarly require extensive social changes. These are likely to be spread over decades rather than years.

The adoption of many new electronic information systems, such as tele-shopping and tele-banking, will depend on institutional and legal changes, and other major government decisions in regional telecommunications investment. The social and political climate in particular countries at particular times may or may not be especially hospitable to these types of social and organizational change. The capacity for social innovation is very variable, and in addition to the capacity to generate and launch a particular group of technical innovations this must surely be one of the main reasons for the changing locus of technological leadership in the various long waves, to which Ray (1980) has so rightly drawn attention.

The capacity for social change and innovation was rather high in the UK in the eighteenth and nineteenth centuries (although bought at a price in working class suffering whose aftermath is still with us). The development of the first major railway network in the world involved not just the invention and innovation of the railway locomotive (which had occurred long before), but a whole series of other inventions and innovations affecting the mechanical engineering industries and the iron and steel industries, as well as necessitating changes in the capital market, in legislation and in the training of a skilled labour force. It seems that the capacity of the Japanese to introduce applications innovations and the related social and organizational changes is (for very different reasons) rather high in relation to micro-electronics, robotics and bio-technologies. This might give them a leading

---

[2] *Wolff's Law:* Wolff was a German economist who in 1912 published four 'laws of retardation of progress'. Essentially, he argued that the scope for improvement in any technology is limited, and that the cost of incremental improvement increases as the technology approaches its long-run performance level. Widely referred to in the retardation literature of the 1930s (see, e.g., S. Kuznets, *Secular Movements in Production and Prices*, Houghton Miflin, 1930, Chapter 2).

role in any new upswing of the world economy even if they have not made the majority of basic innovations which may be associated with that upsurge.

A good example of these points is the automobile industry and Klein (1977) has a nice chapter in his book on *Dynamic Economics* which illustrates most of them. The US did not lead Europe in the earlier stages of the innovation of the internal combustion engine or the passenger car—rather the reverse in fact. Indeed, it was by no means clear that the internal combustion engine would be preferred to the steam engine or electric engine. The basic innovation for all three had been made well before the turn of the century, although the actual dates are still a matter of controversy. Klein points out that in 1900 steam and electric vehicles accounted for 'about three fourths of the four thousand automobiles estimated to have been produced by fifty-seven American firms' (1977, p. 91). The decisive step which the US firms took (as a result of the competitive pressures within the industry) was to reduce the cost of manufacture of the gasoline engine car by more than 50 per cent within a few years. The price of Model T fell from $850 in 1908 to $360 in 1916, sales increased by a factor of 50, market share increased from 10 per cent in 1909 to 60 per cent in 1921, profits on net worth were sometimes as high as 300 per cent per annum and the US attained a dominant position in world export markets. This was indeed 'fast history' analogous to the tempestuous growth of the semi-conductor industry half a century later—with its similar drastic price reductions, rapid changes in market shares, sudden profits for innovating firms and world export hegemony for the leading country until imitators caught up.

The 'basic innovation' which enabled Ford to achieve these dramatic results was, of course, assembly line production. The date of this innovation was right in the middle of the long-wave upswing, not in the earlier depression. In one sense it was a purely organizational innovation, but it both entailed and stimulated a great deal of technical innovation:

> . . . once the organisational change was made, the automobile firms found many opportunities for developing more efficient machines by making them more automatic. For example, replacement of the vertical turret lathe by a more automatic horizontal lathe doubled output per worker. Or to take a more spectacular example, an automatic machine for making camshafts increased output per worker by a factor of ten, and literally dozens of cases can be found in which better machines. permitted output per worker to increase by a factor of between two and ten.

Incidentally, although strongly Schumpeterian in the main thrust of his argument, Klein casts some doubt on the 'heroic' entrepreneur idea so far as it applied to Ford, quoting the comment of Nevins and Hill (1954):

> It is clear that the impression given in *My Life and Work* that the key ideas of mass production percolated from the top of the factory downwards is erroneous: rather seminal ideas moved from the bottom upwards. To be sure, Ford took a special interest in the magneto assembly. But elsewhere able employees like Gregory, Klann and Purdy made important suggestions, Sorensen and others helped them work out, while Ford gave encouragement and counsel. The largest single role in developing the new system, however, was played by the university-trained thinker [Avery] so recently brought in from his school-room.

As we saw in chapters 2, 3 and 4, the linkages with university research were very much stronger in most of the technologies associated with the fourth Kondratiev, but once the industrial application of a new technology begins to develop, to a considerable extent it has its own momentum.

The expression 'natural trajectory' has been coined to describe this process of cumulative exploitation of new ideas. In their paper 'In search of useful theory of innovation', Nelson and Winter (1977) distinguish various types of natural trajectory, including some which are specific to a particular industry or product and some which are of very great general importance such as mechanization. To the best of our knowledge these authors have not attempted directly to relate these ideas to 'long waves' in the economy as a whole but they do point out that:

> . . . there is no reason to believe (and many reasons to doubt) that the powerful general trajectories of one era are the powerful ones of the next. For example, it seems apparent that in the 20th century two widely used natural trajectories opened up (and later variegated) that were not available earlier: the exploitaton of understanding of electricity and the resulting creation and improvement of electrical and later electronic components, and similar developments regarding chemical technologies . . . it is apparent that industries differ significantly in the extent to which they can exploit the prevailing general natural trajectories, and these differences influence the rise and fall of different industries and technologies.

Such general processes of technical change as mechanization, electrification or automation obviously continue over a century or more, but as Coombs (1981) has suggested, the main thrust of their application may be changing, for example, from transfer applications to control applications. Nelson and Winter point out that these very general trajectories of technology are associated strongly with the exploitation of economies of scale. One could therefore expect that, as an industry enters upon a period of rapid expansion, fairly intense efforts will be made to promote innovations which facilitate the attainment of these economies—as was found in chemical process plant in the 1960s and in the semi-conductor industry in the 1950s and 1960s (chapters 2 and 4). The example of the automobile industry in the US also illustrates the point rather nicely.

Secondly, although the search for factor-saving technical change is a constant feature of the innovative efforts of many industries, it is likely to be particularly intense in certain circumstances. One obvious case is the contemporary emphasis on energy-saving technical innovations, apparently increasingly effective in the case of oil, although with a time lag which is directly relevant to the discussion. Another case is the desire to save on imported natural materials in the event of war or threat of war. In the case of labour costs it seems possible that the pressure for labour-saving technical change will be at its most intense during periods of labour shortages and a steep relative rise in labour costs (i.e. at peak periods of prosperity). However, because of the time lags involved in any such change of emphasis and the independent application of such innovations, the actual shedding of labour may occur somewhat later.

New technology, employment and long waves

In the major boom periods new industries and technological systems tend to generate a great deal of new employment, as the form that expansion takes is the installation of completely new capacity and the building up of associated capital goods industries. In the early stages, since the technology is still in a relatively fluid state and standardized special plant and machinery is not yet available, the new factories and plants are often fairly labour-intensive. New small firms may also play an important role among the new entrants and they tend to have a lower than average capital intensity.

However, as a new industry or technology matures, several factors are interacting to reduce the employment generated per unit of investment. Economies of scale become increasingly important and these work in combination with technical changes and organizational changes associated with increasing standardization (Abernathy and Utterback, 1977). The profits of innovation are diminished both by competition and by the pressures on input costs, especially labour costs. A process of concentration tends to occur and competition forces increasing attention to the problem of cost-reducing technical change. This tendency plays an important part in the cyclical movement from boom to recession (or stagflation) and from recession to depression.

Here the Marxist tradition in economics and particularly Mandel (1972) have made an important contribution to the debate in their emphasis on the significance of the general rate of profit and the tendencies which may lead this rate to fall. However, Mandel (and Kuznets) were incorrect in asserting that Schumpeter's theory of the long waves was dependent on a *deus ex machina* of waves of entrepreneurial energy and ignored the importance of profit. The role of profit was built into Schumpeter's theoretical system by definition and he saw innovation as the source of a new range of profit opportunities. However, the 'competing away' of profits during the diffusion of new technologies may well be complemented by the process which Mandel emphasizes—the tendency for the rate of profit to fall throughout the system for reasons associated with the growth of capital intensity and because of greater bargaining strength of labour following long periods of full employment, and related inflationary pressures and social changes.

In the fourth Kondratiev, in particular, it may be that the firms involved in promoting the new wave of technological systems reflect the more oligopolistic nature of capitalism generally and have been better able to delay the 'competing away' of innovation profits. Schumpeter already commented on this tendency in the 1920s, and the strong inflationary pressures during the recession of the fourth Kondratiev may be explained to some extent by a combination of this tendency and the stronger pressures on production costs, including energy and material costs.

The combined effect of these cost pressures and the concentration, scale and standardization effects already mentioned will be to focus technical effort increasingly on cost-reducing process innovations rather than new products. The patterns of investment and R & D during a stagflation period will both tend to shift accordingly—the proportions of rationalization and replacement investment will tend to rise and new capacity investment to diminish; more investment will go into machinery and less into new buildings. In this period, too, there may be increasing pressure for 'improvement innovations' and 'product differentiation'. We would not expect basic innovations to stop, but they may tend to slow down and there will be greater pressures favouring basic process and other cost-saving innovations. Walsh *et al.* (1979) has shown that process patents rose much more rapidly than product patents in the plastics industry in the 1950s and 1960s (see chapter 3), and the SPRU data bank on innovations indicates a more general tendency in this direction across many industries.

Finally, the descent from stagnation to depression may be explained by 'overshoot' arising from efforts to combat inflationary pressures but also by the successive weakening of the expansionary elements within the long wave. In particular, the shift in the main thrust of investment and technical development will mean that the investment required to generate further increments of new employment will be rising, so that even when investment is expanded during the successive Juglar cycle upswings as a result of Keynesian or other stimuli, it has less effect in expanding employment, unless it is directed to areas of very low capital intensity such as various government and administrative services.

The role of technology policy

Throughout this book it has been maintained that the emergence of new technologies and their assimilation in the economies of industrial countries, which is an extremely important aspect of long-term economic development, is not a smooth continuous process. Specifically, it has been argued that technical change is extremely uneven over *time*; as between *industries* and broad sectors of the economy; and *geographically* as between regions and countries. The diffusion of clusters of technical innovations of wide adaptability is capable of imparting a substantial upthrust to the growth of the economic system, creating many new opportunities for investment and employment and generating widespread secondary demands for goods and services. Over *time*, however, these new 'technological systems' mature and their investment and employment consequences tend to change. The 'compensation effects', which might mitigate these tendencies, operate only imperfectly and often with long delays. Much of the stock of capital equipment already in existence in manufacturing may be rather inflexible in terms of its employment potential (i.e. it is 'clay' rather than 'putty') and new vintages of equipment may require heavy investment.

The promotion of major new technological systems and of productivity growth based on technical change may be an important means to help restore the economic health of the mature industrialized countries. This conclusion emerges from the earlier experiences of recovery from depression as well as from the more recent Japanese experience. Particularly important would be innovations of wide adaptability, which could lead to a combination of high productivity gains and greatly improved profitability.

Three sets of technology policies seem particularly relevant:

(i) Policies which aim directly at encouraging firms to take up radical inventions/innovations. They seem particularly relevant in those recessional/depressional phases, when private investment seems reluctant to go for these radical, but risky, innovations. They could be described as Schumpeter Mark I policies, ranging from direct financial support, to various forms of indirect risk-taking support in order to promote the emergence of radical new innovations. The evidence that has been considered in this book indeed suggests that such radical innovations are often not immediately and obviously profitable. Only after a fairly long gestation period does 'take-off' occur. This means that during the gestation period positive and patient public policies of support, encouragement, experiment and adaptation can be extremely important. The computer is a particularly notable example. The unaided market mechanism is not enough.

As we saw in chapter 9, economists and a wider public have become suspicious of governments putting public money into exotic new technologies because of disappointing past experiences with supersonic transport aircraft, and some military and nuclear projects. However, these were launched in quite different circumstances and often without much consideration of the economic and social aspects.

Exploratory development would often merit support from government sources, but full-scale commercial development, which is usually far more expensive, more seldom justifies the commitment of public funds to R & D. The expensive failures that several countries have experienced have largely stemmed from disregard of this basic distinction, from the power and prestige of specialized lobbies to influence government policies and from the associated absence of adequate public debate. Investment projects incorporating new equipment, and procurement of new products that meet advanced technical specifications and satisfy social requirements, may be a much more satisfactory form of public involvement at this stage than R & D subsidies.

There are many other ways in which public policies can help to promote the emergence of radical new technologies. Some of them are described in general terms in the OECD Report on *Technical Change and Economic Policy* (1980) and by Rothwell and Zegveld (1981). Support for fundamental research, for 'enabling' or 'fundamental' technology and policies designed to improve the coupling system are all of critical importance.

(ii) However, the early stages of radical innovations do not have big economic effects. Only large-scale diffusion can have such effects and therefore a second set of policies aimed at improving the diffusion of existing, but still relatively new and radical, innovations throughout the various sectors is essential.

Moreover, such ambitious long-term policies should pay special attention to the needs of the education system, the health services and other social services in which direct public procurement and investment are essential, and which might otherwise be the Cinderella areas of social and technical change.

The skill shortages that inhibit the rapid diffusion of new technology systems are an obvious and important area of public involvement. Experience in Germany and Sweden in particular has demonstrated that imaginative and ambitious public programmes can complement the efforts of private firms in training and re-training the labour force, and minimize the adjustment problems of structural change especially for young people. Experience in Japan suggests that massive public investment in higher education combined with intensive training and re-training in industry may be even more effective. A much higher proportion of young Japanese attend higher education than in any European country now. There is scope for a variety of strategies tailored to national circumstances but there can be little doubt that a policy of 'intangible investment' in the expansion of education and training can be a useful aid to adaptation and a valuable 'counter-cyclical' investment strategy.

(iii) A third set of policies aims at improving the import and the internal diffusion of foreign technology. It is a policy that in the first instance has to convince local businessmen and managers, as well as government officials, that foreign technology in certain areas and at certain times might be more advanced or simply better than domestic technology; this seems to be particularly difficult to achieve in the case of the old technological leaders, such as the UK and the US. Yet, as Japanese post-war experience shows, a deliberate policy towards the import of foreign technology, coupled with autonomous efforts to improve it, can be highly successful. Particularly for industries that are at some distance from the world technological frontier, such a policy seems extremely relevant, but even for technological leaders, active support in seeking and using the best available world technology is common sense. No country can be a technological leader in all areas, and all can learn from international experience. Perhaps the most important group in MITI is that which monitors world-wide technological developments and advises on possible future trends and their implications for Japanese industry.

The limitations of technology policy and of demand management

It would be a mistake to believe that technology and training policies *alone*, however well conceived and executed, could extricate market economies from their present difficulties. Technology policy does not operate in a vacuum but in a very specific economic and political environment. The co-existence of high inflation with high unemployment is a major new phenomenon of the present Kondratiev downswing. Chesnais (1982) has reminded us that each successive wave of new technologies is diffusing in a very different social climate; Schumpeter discussed the new complications arising from the growing role of the state and of large oligopolies already in the recovery from the 1930s depression. He returned to these problems particularly in *Capitalism, Socialism and Democracy* (1942).

Schumpeter Mark I is still important but Schumpeter Mark II predominates. The inflexibility of wages has been a commonplace of the economic policy debate for more than half a century. There is much force in Kristensen's (1981) argument that almost all markets, including commodities and capital as well as labour, have become much less flexible than in the nineteenth century or the early part of this century. His analysis of inflation and unemployment is one of the most original and challenging in the recent debate and derives added force from his experience as a Finance Minister and as Secretary General of the OECD in the 1960s. After discussing the loss of price flexibility in many types of 'organized' market, he concludes:

> ... expansionary demand management may be able only temporarily to influence employment in a positive direction, and . . . the chances of lowering the rate of inflation through restrictive policies are poor. The possibilities of getting out of the present unhappy combination of inflation and unemployment must therefore be considered small or non-existent (p. 43).

He goes on to ask whether in these circumstances there is (as some monetarists suggest) a 'natural' and unavoidable rate of unemployment or of inflation and answers his own question:

> In my view the word 'natural' is misleading in both cases. It is not in the nature of things that there must always be high rates of unemployment or of inflation. Rather with the present organisation of markets and of the public sector a strong tendency for both inflation and unemployment to be persistent features of modern economies seems to be built into the system. This does not prevent the

performance of the system from being better in both respects if these systems of organisation are improved (p. 43).

In the concluding chapter of his book (in which he discusses the policy implications of his analysis), he raises the other fundamental question which is at the heart of the current debate:

If it is true that the economies have lost much of the self-adjusting capacity of former times, can we not go back to the old systems of the nineteenth century, when prices, wages and employment moved up and down but never got locked in an extreme situation for any length of time?

Again he gives an unequivocal answer:

There is no possibility of moving backward in history. Concentration of industry has been unavoidable because large enterprises are superior in branches where certain overhead costs, such as research and development, are important. The larger the market, the lower these costs per unit produced. For the same reasons multi-national corporations and international banking operations were bound to expand. . . . Similarly, it could not be avoided that wage earners united more and more. . . . Finally the public sector has been bound to expand in modern societies. . . . If we cannot go backward we must go forward. We must accept that markets are organised and increasingly so. And we must accept that the political authorities of national states have important roles to play, now supplemented by a number of international organisations (p. 134).

He concludes that the performance of the OECD economies can only be improved by the institutionalization of prices and incomes policies on a semi-permanent basis. Only such radical new departures could ensure a return to price stability and full employment.

Obviously, not everyone has been convinced by Kristensen's analysis and his powerful restatement of the radical Keynesian alternative to monetarism. But in so far as it relates to the role of innovation in the concentration process, to the growth of inflexibilities in markets, and to the role of government in policies for science and technology, his arguments are persuasive. Whether economies which adopted his policy prescriptions could any longer be reasonably described as 'market' economies is an open question, but in any case we have already made clear that our approach to the behaviour of the economic system, like that of Schumpeter and Kristensen, is essentially a historical one which tries to take account of institutional and social change, as well as the specific features of successive waves of new technology.

Clearly a zero-growth economy, or still more a declining economy, cannot satisfy the aspirations of its citizens for a rising standard of living or can only satisfy the ambitions of a small fraction of them at the expense of the rest, with all the potential for social conflict which that implies. A rapidly growing economy, on the other hand, risks generating inflationary pressures which may become unacceptable if a 'free' labour market operates in conditions of strong demand for labour and acute skill shortages.

Whether or not the alternative solution of some kind of incomes policy can be made to 'stick' depends, in part, on the social learning process that we have gone through in the past quarter of a century. However, it would at least be more likely to stick if part of the deal were a significant, though small, improvement in the real income of the majority of the population on a relatively regular basis. Such an improvement is possible only through a high rate of technical change. If the responsibility for that technical change and the rewards for its effective implementation are more widely diffused and the whole process is more widely understood and appreciated, then this can only benefit the implementation of growth-oriented but anti-inflationary policies. 'Self-management', 'worker participation' and co-operative ownership are all important in this context.

To combat inflationary pressures with deflationary policies and high unemployment puts many of the social and political achievements of the entire post-war period at risk.

The problem of unemployment is the most serious confronting the industrialized countries in the 1980s and it must always be remembered that Hitler's accession to power in 1933 was directly associated with the effects of the Great Depression. Support for his party remained relatively small, even immediately after the runaway inflation of 1923, but his vote soared from 1930 to 1932, when mass unemployment provided a fertile breeding ground for his doctrines and thousands of recruits for the storm troops and other paramilitary formations. The riots in English cities in 1981 were a further salutary reminder of the potentially explosive psychological and political consequences of large-scale unemployment associated with inner-city decay.

Jahoda (1982) has concluded that the destructive psychological consequences of unemployment for the individual are more severe than the alienating effects of the less pleasant forms of employment. She has also concluded that the harmful effects of unemployment in the 1980s are potentially as severe as those of the 1930s, in spite of the intervening changes in social welfare arrangements. She was one of the first to study the destructive social psychological consequences of prolonged unemployment in Marienthal, a village in Austria, in the 1930s and her views must command great respect.

The acceptance of a high level of unemployment as a form of restraint on wage pressures, whether temporary or permanent, is socially unacceptable and politically impracticable over any extended period. There has to be a better way to treat human beings in the twentieth-century. I am not suggesting that technology policies alone can be successful in solving the fundamental social, political and economic problems that go far beyond the scope of this book. However, well-conceived technology policies are a vital ingredient of any strategy designed to combat the twin crises of unemployment and inflation.

APPENDIX

Excerpt from: *The Measurement of Scientific and Technical Activities: Proposed Standard Practice for Surveys of Research and Experimental Development ('Frascati Manual'), 1980,* OECD, Paris, 1981, Chapter II, pp. 25–37.

Chapter II

BASIC DEFINITIONS AND CONVENTIONS

2.1 RESEARCH AND EXPERIMENTAL DEVELOPMENT (R & D)

43.   Research and experimental development (R & D) comprise creative work undertaken on a systematic basis in order to increase the stock of knowledge, including knowledge of man, culture and society and the use of this stock of knowledge to devise new applications.

R & D is a term covering three activities: basic research, applied research and experimental development.[1] *Basic research* is experimental or theoretical work undertaken primarily to acquire new knowledge of the underlying foundation of phenomena and observable facts, without any particular application or use in view. *Applied research* is also original investigation undertaken in order to acquire new knowledge. It is, however, directed primarily towards a specific practical aim or objective. *Experimental development* is systematic work, drawing on existing knowledge gained from research and/or practical experience that is directed to producing new materials, products or devices, to installing new processes, systems and services, or to improving substantially those already produced or installed.

2.2 ACTIVITIES TO BE EXCLUDED FROM R & D

44.   For survey purposes R & D must be distinguished from a wide range of related activities with a scientific and technological base. These other activities are very closely linked to R & D through flows of information and in terms of operations, institutions and personnel, but they should, as far as possible, be excluded when measuring R & D.

45.   These activities will be discussed here under three headings:
   — Education and training (see 2.2.1);
   — Other related scientific and technological activities (see 2.2.2);
   — Other industrial activities (see 2.2.3).
The definitions here are practical and designed solely to exclude these activities from R & D. They are thus slightly different from the broader concepts of "STET", "STS" and "innovation" discussed in Chapter I.

2.2.1 *Education and Training*

46.   All education and training of personnel in the natural sciences, engineering, medicine, agriculture, the social sciences and the humanities in universities and special institutions of higher and post-secondary education. However, bona fide research by postgraduate students carried out at universities should be counted, wherever possible, as a part of R & D (see also 2.3.2).

---

[1] Described in detail in Chapter IV of the publication from which this extract is taken.

2.2.2 *Other Related Scientific and Technological Activities*

47.   The following activities should be excluded from R & D except where carried out solely or primarily for the purpose of an R & D project (see also examples in section 2.3.1):

2.2.2.1 *Scientific and Technical Information Services*

48.   The specialised activities of:

— collecting — coding — recording — classifying — disseminating — translating — analysing — evaluating	by	— scientific and technical personnel — bibliographic services — patent services — scientific and technical information extension and advisory services — scientific conferences

except where conducted solely or primarily for the purpose of R & D support (e.g. the preparation of the original report of R & D findings should be included in R & D).

2.2.2.2 *General Purpose Data Collection*

49.   Undertaken generally by government agencies to record natural biological or social phenomena that are of general public interest or that only the government has the resources to record. Examples are routine topographical mapping, routine geological, hydrological, oceanographic and meteorological surveying, astronomical observations. Data collection conducted solely or primarily as part of the R & D process is included in R & D (e.g. data on the paths and characteristics of particles in a nuclear reactor). The same reasoning applies to the processing and interpretation of the data. The social sciences, in particular, are very dependent on the accurate record of facts relating to society in the form of censuses, sample surveys, etc. When these are specially collected or processed for the purpose of scientific research the cost should be attributed to research and should cover the planning, systematising etc. of the data. But data collected for other or general purposes such as quarterly sampling of unemployment, should be excluded even if exploited for research. Market surveys are excluded.

2.2.2.3 *Testing and Standardization*

50.   Refers to the maintenance of national standards, the calibration of secondary standards and routine testing and analysis of materials, components, products, processes, soils, atmospheres, etc.

2.2.2.4 *Feasibility Studies*

51.   Investigation of proposed engineering projects using existing techniques in order to provide additional information before deciding on implementation. In the social sciences, feasibility studies are investigations of the socio-economic characteristics and implications of specific situations (e.g. a study of the viability of a petro-chemical complex in a certain region). However, feasibility studies on research projects are part of R & D.

2.2.2.5 *Specialised Medical Care*

52.   Refers to routine investigation and normal application of specialised medical knowledge. There may, however, be an element of R & D in what is usually called "advanced medical care", carried out, for example, in university hospitals.

2.2.2.6 *Patent and Licence Work*

53.   All administrative and legal work connected with patents and licences. (However, patent work connected *directly* with R & D projects is R & D).

2.2.2.7 *Policy Related Studies*

54.   Policy in this context refers not only to national policy but also to policy at the regional and local levels, as well as that of business enterprise in the pursuit of their economic activity. Policy related studies cover a range of activities such as the analysis and assessment of the existing programmes, policies and operations of government departments and other institutions; the work of units concerned with the continuing analysis and monitoring of external phenomena (e.g. defence and security analysis); and the work of legislative commissions of inquiry concerned with general government or departmental policy or operations.

2.2.3 *Other Industrial Activities*

55.   These can be considered under two, to some extent overlapping, headings:

2.2.3.1 *Industrial Innovation (not elsewhere classified)*

56.   All those scientific, technical, commercial and financial steps, other than R & D, necessary for the successful development and marketing of a manufactured product and the commercial use of the processes and equipment.*

2.2.3.2 *Production and Related Technical Activities*

57.   Industrial production and distribution of goods and services and the various allied technical services in the Business Enterprise sector and in the economy at large, together with allied activities using the disciplines of the social sciences such as market research.

2.3 THE BOUNDARIES OF R & D

2.3.1 *The Basic Criterion*

58.   The basic criterion for distinguishing R & D from related activities is the presence in R & D of an appreciable element of novelty.

(Supplementary criteria are suggested in Chapter VI—see 6.3.3). One aspect of this criterion is that a particular project may be R & D if undertaken for one reason but if carried out for another reason will not be considered R & D. This is shown in the following examples:

(a)   In the field of medicine, routine autopsy on the causes of death is simply the practice of medical care and *not* R & D; special investigation of a particular mortality in order to establish the side effects of certain cancer treatment *is* R & D. Similarly routine tests, such as blood and bacteriological tests carried out for doctors, are *not* R & D but a special programme of blood tests in connection with the introduction of a new drug *is* R & D.

(b) The keeping of daily records of temperatures or of atmospheric pressure is *not* R & D but the operation of a weather forecasting service or general data collection. The investigation of new methods of measuring temperature *is* R & D, as are the study and development of new systems and techniques for interpreting the data.

(c) R & D activities in the mechanical engineering industry often have a close connection with design and drawing work. Usually there are no special R & D departments in small and medium size companies in this industry and R & D problems are mostly dealt with under the general heading "design and drawing". If calculations, designs, workshop drawing and operating instructions are made for the setting-up and operating of pilot plants and prototypes, they should be included in R & D. If they are carried out for the preparation, execution and maintenance of production standardization (e.g. jigs, machine tools) or to promote the sale of products (e.g. offers, leaflets, spare parts catalogues) they should be excluded from R & D.

(d) Many social scientists perform work in which they bring established methodologies and facts of the social sciences to bear on a particular problem, but which cannot be classified as research. The following examples of work which might come in this category and are not R & D: interpretative commentary on the probable economic effects of a change in the tax structure, using existing economic data; forecasting future changes in the patterns of the demand for social services within a given area arising from an altered demographical structure; operations research (OR) as a contribution to decision making, e.g. planning the optimal distribution system for a factory; the use of standard techniques in applied psychology to select and classify industrial and military personnel, students, etc. and to test children with reading or other disabilities.

2.3.2 *Problems at the Borderline Between R & D, and Education & Training*

2.3.2.1 *General Approach*

59. In institutions of higher education, research and teaching are always very closely linked, as most academic staff do both and many buildings, as well as much equipment, serve both purposes. In the absence of complete and accurate information, measurement of the share of R & D is generally based on estimates of the proportion of working time devoted to this activity by university staff. This is a very important estimate, especially in the social sciences and humanities where a particularly high proportion of research is carried out in the universities.

2.3.2.2 *The Case of Postgraduate Studies*

60. The borderline between R & D and Education and Training is particularly hard to establish in the case of postgraduate education (i.e. at ISCED* level category 7) which involves training in research. The activities of both the postgraduate students themselves and of their teachers need to be taken into consideration.

61. Parts of the curricula for postgraduate studies (ISCED level category 7) are highly structured, involving, for instance, study schemes, set courses, compulsory laboratory work, etc. Here, the teacher is disseminating education and training in research methods. Typical activities for students under this heading

are attending compulsory courses, studying literature on the subject concerned, learning research methodology, etc. These activities do not fulfil the criterion of novelty specified in the definition of R & D.

62. In addition, in order to obtain a final qualification at postgraduate level (ISCED 7) students are also expected to prove their competence by undertaking a relatively independent study or project and by presenting its results. As a general rule, these studies contain the elements of novelty required for R & D projects. The relevant activities of such students should, therefore, be attributed to R & D, any supervision by the teacher should also be considered as R & D. In addition to R & D performed within the framework of courses of postgraduate education, it is possible for both teachers and students to be engaged on other R & D projects.

63. Finally, such students at this level are often attached to or directly employed by the establishment concerned and have contracts or are bound by a similar engagement, which oblige them to do some teaching at lower levels or to perform other activities such as advanced medical care whilst allowing them to continue their studies and to do research.

64. The borderline between R & D and Education at ISCED level 7 are illustrated in Table II.1 which, together with much of the above text, is based on the Nordic Manual "Statistics of Resources Devoted to Higher Education". The more practical problems of applying these concepts are dealt with in Chapter V (see 5.2.2.2).

2.3.3 *Problems at the Borderline between R & D and Other Related Scientific and Technological Activities*

2.3.3.1 *General Approach*

65. Difficulties in the separation of R & D from other scientific and technological activities are caused by the performance of several activities at the same institution. In survey practice the identification of the R & D portion is facilitated by certain rules of thumb. Two typical illustrations of the use of these may be cited:

— Institutions or units of institutions and firms whose principal activity is R & D often have secondary, non-R & D activities (e.g. scientific and technical information, testing, quality control, analysis). In so far as a secondary activity is undertaken primarily in the interests of R & D, it should be included in R & D activities; if the secondary activity is designed essentially to meet needs other than R & D, it should be excluded from R & D.

— Institutions whose main purpose is an R & D related scientific activity often undertake some research in connection with this activity. Such research should be isolated and included when measuring R & D.

Examples

(a) The activities of a scientific and technical information service or of a research laboratory library, maintained predominantly for the benefit of the research workers in the laboratory, should be included in R & D. The activities of a firm's documentation centre open to all the firm's staff should be excluded from R & D even if it shares the same premises as the company research unit. Similarly, the activities of central university libraries should be excluded from R & D.

These criteria apply only to the cases where it is necessary to exclude the activities of an institution or a department in their entirety. Where more detailed accounting methods are used it may be possible to impute part of the costs of the excluded activities as R & D overheads. Whereas the preparation of scientific and technical publications is, generally

Table II.1   Borderline between R & D and education and training at ISCED national level category 7

	Education & Training at Level 7	R & D	Other Activities
Teachers	1. Teaching students at level 7 2. Training students at level 7 in R & D methodology, laboratory work, etc.		
		3. Supervision of R & D projects required for students' qualification at level 7 4. Supervision of other R & D projects and performance of own R & D projects.	
			5. Teaching at levels lower than 7. 6. Other activities.
Post-graduate Students	1. Courses work for formal qualification including independent study, work etc.		
		2. Performing and writing up R & D projects required for formal qualification. 3. Any other R & D activities.	
			4. Teaching at levels lower than 7. 5. Other activities.

    speaking, excluded, the preparation of the original report of research findings should be included in R & D.

(b) Public bodies and consumer organisations often operate laboratories where the main purpose is testing and standardization. The staff of these laboratories may also spend time devising new or substantially improved methods of testing. Such activities should be included in R & D.

(c) General purpose data collection is particularly important to social science research, since without it many elements of this research would not be feasible. However, unless it is collected primarily for research purposes, it should not be classified as a research activity. On the other hand, the larger statistical institutes may carry out some R & D (e.g. on survey methods, sampling methodologies and small area statistical estimates). Whenever possible, such R & D should be identified and appropriate estimates included with the main R & D sectoral data.

2.3.3.2 *Specific Cases*

66.   In certain cases the theoretical criteria for distinguishing between R & D and related technological activities are particularly difficult to apply. Space exploration

and mining and prospecting are two areas where large amounts of resources are involved and so any variations in the way they are treated will have important effects on the international comparability of the resulting R & D data. For this reason, the following conventions apply in these two cases:

2.3.3.1.1 *Space Exploration*

67. The difficulty with space exploration is that, in some respects, much space activity may now be considered routine; certainly the bulk of the costs are incurred for the purchase of goods and services which are not R & D. However, the object of all space exploration is still to increase the stock of knowledge so that it should all be *included* in R & D. It may be necessary to separate these activities associated with space *exploration*, including the development of vehicles, equipment and techniques, from those involved in the routine placing of orbiting satellites or establishment of tracking and communication stations.

2.3.3.2.2 *Mining and Prospecting*

68. Mining and prospecting sometimes cause problems due to a linguistic confusion between "research" for new or substantially improved resources (food, energy, etc.) and the "search" for existing reserves of natural resources which blurs the distinction between R & D and surveying and prospecting. In theory in order to establish accurate R & D data, the following activities should be identified, measured and summed:

(i) The development of new surveying methods and techniques.
(ii) Surveying undertaken as an integral part of a research project on geological phenomena.
(iii) Research on geological phenomena *per se* undertaken as a subsidiary part of surveying and prospecting programmes.

In practice, the third of these presents a number of problems. It is difficult to frame a precise definition which would be meaningful for respondents to national surveys. The sums involved are probably relatively small in practice but a misreading by respondents might lead to large amounts of "search" resources being counted as R & D. For this reason, only the following activities should be included in R & D:

— The development of new or substantially improved methods and equipment for data acquisition and for the processing and study of the data collected and for the interpretation of these data;
— Surveying undertaken as an integral part of an R & D project on geological phenomena *per se* including data acquisition, processing and interpretation undertaken for primarily scientific purposes.

It follows that the surveying and prospecting activities of commercial companies will be almost entirely excluded from R & D. For example, the sinking of exploratory wells to evaluate the resources of a deposit should be considered as scientific and technological services.

2.3.4 *Problems on the Borderline Between R & D and Other Industrial Activities* (see also Table II.2)

2.3.4.1 *General Approach*

69. Care must be taken to exclude activities which, though undoubtedly a part of the innovation process, rarely involve any R & D, e.g. design engineering, patent filing and licensing, "tooling up" and market research. Similar difficulties may arise in distinguishing public technology based services such as inspection and control from related R & D, as for example in the area of foods and drugs.

70.   A precise definition of the cut-off point between experimental development and production cannot be stated in such a way that it is applicable to all industrial situations—instead, it would be necessary to establish a series of conventions or criteria by type of industry. However, the basic rule laid down by the National Science Foundation (NSF) provides a practical basis for the exercise of judgement in difficult cases. Slightly expanded, it states:

> "If the primary objective is to make further technical improvements on the product or process, then the work comes within the definition of R & D. If, on the other hand, the product, process or approach is substantially set and the primary objective is to develop markets, to do preproduction planning or to get a production or control system working smoothly, then the work is no longer R & D."

2.3.4.2 *Specific Cases*
71.   Some common problem areas are described below:
2.3.4.2.1 *Prototypes*
72.   A prototype is an original model on which something new is patterned and of which all things of the same type are representations or copies. It is a basic model possessing the essential characteristics of the intended product. Applying the NSF criterion, the design, construction and testing of prototypes *normally* falls within the scope of R & D. This applies whether only one or several prototypes are made and whether consecutively or simultaneously. But when any necessary modifications to the prototype(s) have been made and testing has been satisfactorily completed, the boundary of R & D has been reached. The construction of several copies of a prototype to meet a temporary commercial, military or medical need after successful testing of the original, even if undertaken by R & D staff, is not part of R & D.

2.3.4.2.2 *Pilot Plants*
73.   The construction and operation of a pilot plant is a part of R & D as long as the principal purposes are to obtain experience and to compile engineering and other data to be used in:
   — evaluating hypotheses;
   — writing new product formulae;
   — establishing new finished product specifications;
   — designing special equipment and structures required by a new
       process;
   — preparing operating instructions or manuals on the process.

But if, as soon as this experimental phase is over, a pilot plant switches to operating as a normal commercial production unit, the activity can no longer be considered R & D even though it may still be described as "pilot plant". As long as the primary purpose in operating a pilot plant is non-commercial, it makes no difference in principle if part or all of the output happens to be sold. Receipts from this source should not be deducted from the cost of R & D activity. However, as soon as pilot plant begins to operate as a normal production unit, the effect is more or less the same as the sale of a pilot plant.

2.3.4.2.3 *Very Costly Pilot Plants and Prototypes*
74.   It is very important to look closely at the nature of very costly pilot plants or prototypes, for example the first of a new line of nuclear power stations or of ice-breakers. They may be constructed almost entirely from existing materials and

using existing technology and they are often built for use simultaneously for R & D and to provide the primary service concerned (power generation or ice-breaking). The construction of such plants and prototypes should not be wholly credited to R & D. For further details see Chapter V [5.3.2.3.4 and 5.3.3.2.2 (ii) and (ii)].

2.3.4.2.4 *Trial Production*

75.　After a prototype, with any necessary modifications, has been satisfactorily tested, the costs of the first trial production runs should not be attributed to R & D since the primary objective is no longer to make further improvements to the product but to get the production process going. The first units of a trial production run for a mass production series should not be considered as R & D prototypes, even if they are loosely described as such. Normally, the costs of trial product runs of "experimental production", including tooling-up for full-scale production (tool making and tool try-out) are not to be included in R & D.

2.3.4.2.5 *Trouble-shooting*

76.　Trouble-shooting occasionally brings out the need for further R & D but more frequently it involves the detection of faults in equipment or processes and results in minor modifications of standard equipment and processes. It should not, therefore, be included in R & D.

2.3.4.2.6 *"Feed-back" R & D*

77.　After a new product or process has been turned over to production units, there will still be technical problems to be solved, some of which may demand further R & D. Such "feedback" R & D should be included.

Table II.2　Some borderline cases between R & D and other industrial activities

Item	*Treatment*	*Remarks*
Prototypes	Include in R & D	As long as the primary objective is to make further improvements.
Pilot plant	Include in R & D	So long as the primary purpose is R & D
Design and drawing	Divide	Include design required during R & D. Exclude design for production process
Trial production and tooling-up	Exclude	Except "feed-back" R & D
After-sales service and trouble-shooting	Exclude	Except "feed-back" R & D
Patent and licence work	Exclude	All administrative and legal work connected with patents and licences. (Except patent work *directly* connected with R & D projects.)
Routine tests	Exclude	Even if undertaken by R & D staff
Data collection	Exclude	Except when an integral part of R & D
Public inspection control, enforcement of standards, regulations	Exclude	

References

Abernathy, W. J. and Utterback, J. M. (1975), 'A dynamic model of process and product innovation', *Omega*, vol. 3, no. 6.

Abernathy, W. J. and Utterback, J. M. (1978), 'Patterns of industrial innovation', *Technology Review*, vol. 80, June–July.

Abernathy, W. J. and Utterback, J. M. (1979), 'Dynamics of innovation in industry', in Hill, C. T., and Utterback, J. M. (eds), *Technological Innovation for a Dynamic Economy*, Pergamon, Oxford.

Achilladelis, B. G. (1973), 'Process Innovation in the Chemical Industry', D.Phil. thesis, University of Sussex.

Allen, D. H. (1968), 'Credibility forecasts and their application to the economic assessment of novel R and D projects', *Operational Research Quarterly*, p. 25.

Allen, D. H. (1972), 'Credibility and the assessment of R and D projects', *Long Range Planning*, vol. 5, no. 2, pp. 53–64.

Allen, G. C. (1981), 'Industrial policy and innovation in Japan', in Carter, C. (ed.), *Industrial Policy and Innovation*, Heinemann, London, pp. 68–87.

Allen, J. A. (1967), *Studies in Innovation in the Steel and Chemical Industries*, Manchester University Press.

Allen, J. M. and Norris, K. P. (1970), 'Project estimates and outcomes in electricity generation research', *J. Manag. Stud.*, vol. 7, no. 3, pp. 271–87.

Allen, T. J. (1966a), 'The performance of information channels in the transfer of technology', *Indust. Manag. Rev.*, vol. 8, no. 1, pp. 87–98.

Allen, T. J. (1966b), *Managing the Flow of Scientific and Technical Information*, Ph.D. thesis, MIT.

Allen, T. J. (1967), 'Communications in the R and D laboratory', *Technology Review*, vol. 70, no. 1, November, pp. 31–7.

Allen, T. J. and Marquis, D. G. (1966), 'Communication patterns in applied technology', *Amer. Psychologist*, vol. 21, pp. 1052–60.

Ames, E. (1961), 'Research, invention, development and innovation', *Amer. Econ. Rev.*, vol. 51, no. 3, pp. 370–81.

Andress, J. F. (1954), 'The learning curve as a production tool', *Harv. Bus. Rev.*, January.

Armytage, W. H. G. (1965), *The Rise of the Technocrats*, Routledge & Kegan Paul.

Arrow, K. J. (1962), 'Economic welfare and the allocation of resources for invention', in National Bureau of Economic Research, *The Rate and Direction of Inventive Activity*, Princeton University Press.

Baker, N. R. and Pound, W. H. (1964), 'R and D project selection: where we stand', *IEEE Trans. Engin. Manag.*, vol. Em-11, no. 4, p. 124.

Baker, R. (1976), *New and Improved—Inventors and Inventions that have Changed the Modern World*, British Museum Publications, London.

Barna, T. (1962), *Investment and Growth Policies in British Industrial Firms*, NIESR Occ. Paper XX, Cambridge University Press.

BASF (1965), *Im Reiche der Chemie*, Econ-Verlag.

Baumler, E. (1968), *A Century of Chemistry*, Econ-Verlag.

Beattie, C. J. and Reader, R. D. (1971), *Quantitative Management in R and D*, Chapman & Hall.

Beer, J. J. (1959), *The Emergence of the German Dye Industry*, University of Illinois Press.

Belden, T. G. and Belden, M. R. (1962), *The Lengthening Shadow: the Life of Thomas J. Watson*, Little Brown, Boston.

Beloff, N. (1966), 'The learning curve—some controversial issues', *J. Int. Econ.*, June.

Ben-David, J. (1968), *Fundamental Research and the Universities: some Comments on International Differences*, OECD.

Benham, F. (1938), *Economics*, Pitman.

Bernal, J. D. (1929), *The World, the Flesh and the Devil: an Inquiry into the Three Enemies of the Rational Soul*, Cape, London.

Bernal, J. D. (1939), *The Social Function of Science*, Routledge & Kegan Paul.

Bernal, J. D. (1958), *World Without War*, Routledge & Kegan Paul.

Braun, E. and MacDonald, S. (1978), *Revolution in Miniature: the History and Impact of Semi-Conductor Electronics,* Cambridge University Press.

Briggs, A. (1961), *The History of Broadcasting in the UK,* Oxford University Press.

Bright, J. F. (ed.) (1968), *Technological Forecasting for Industry and Government,* Prentice-Hall.

Brock, G. W. (1975), *The US Computer Industry—a Study of Market Power,* Cambridge, Mass.

Burke, F. E. (1970), 'Logic and variety in innovation processes', in M. Goldsmith (ed.), *Technological Innovation and the Economy,* Wiley.

Burn, D. L. (1967), *The Political Economy of Nuclear Energy,* Institute of Economic Affairs, London.

Burns, T. and Stalker, G. (1961), *The Management of Innovation,* Tavistock.

Carter, C. (ed.) (1981), *Industrial Policy and Innovation,* Heinemann, London.

Carter, C. F. and Williams, B. R. (1957), *Industry and Technical Progress,* Oxford University Press.

Carter, C. F. and Williams, B. R. (1959a), 'The characteristics of technically progressive firms', *J. Ind. Econ.,* vol. 7, no. 2, pp. 87–104.

Carter, C. F. and Williams, B. R. (1959b), *Science in Industry,* Oxford University Press.

Carter, C. F. and Williams, B. R. (1964), 'Government scientific policy and the growth of the British economy', *Minerva,* vol. 3, no. 1, pp. 114–25.

Centre for the Study of Industrial Innovation (1971), *On the Shelf: a Survey of Industrial R and D Projects Abandoned for Non-technical Reasons.*

Centre for the Study of Industrial Innovation (1972), *Success and Failure in Industrial Innovation,* February.

Chesnais, F. (1982), 'Schumpeterian recovery and the Schumpeterian perspective', in Giersch, H. (ed.) (1982).

Clark, J. A., Freeman, C., and Soete, L. L. G. (1981), 'Long waves, inventions and innovations', *Futures,* vol. 13, no. 4, pp. 308–22.

Clark, R. (1979), *The Japanese Company,* Yale University Press.

Cole, H. S. D. *et al.* (1973), *Thinking about the Future,* Chatto & Windus, London.

Coombs, R. W. (1981), 'Innovations, automation and the long wave theory', *Futures,* vol. 13, no. 5, pp. 360–70.

Cooper, C. M. (1973), 'Choice of techniques and technological change as problems in political economy', *Int. Soc. Sci. J.,* vol. XXV, no. 3, pp. 293–304.

Cooper, C. M. and Clark, J. A. (1982), *Investment, Technical Change and the Level of Employment,* Wheatsheaf, Brighton.

Cooper, C. M. and Sercovich, F. (1971), *Mechanisms for the Transfer of Technology from Advanced to Developing Countries,* UNCTAD, Geneva.

Cooray, D. V. B. N. (1980), 'The technological factor and its relevance to the competition between natural and synthetic rubber in international trade', D.Phil. thesis, University of Sussex.

Cottrell, A. (1981), *How Safe is Nuclear Energy?,* Heinemann, London.

Davies, G. B. (1972), 'Contingency planning: the neglected end of the planning cycle', *Process Engineering,* December, pp. 93–7.

Davies, S. (1979), *The Diffusion of Process Innovations,* Cambridge University Press.

De Bell, J. M. (1946), *German Plastics Practice,* Springfield.

Dean, B. V. (1968), *Evaluating, Selecting and Controlling R and D Projects,* American Management Association, p. 49.

Dedijer, S. (1964), 'International comparisons of science', *New Scientist,* vol. 21, no. 379, pp. 461–4.

Delorme, J. (1962), *Anthologie des Brevets sur les Matières Plastiques,* vols. 1–3, Amphora.

Denison, E. (1962), *The Sources of Economic Growth in the United States and the Alternatives Before us,* Allen & Unwin, London.

Diebold, J. (1952), *Automation: the Advent of the Automatic Factory,* New York, van Nostrand.

Dory, J. P. and Lord, R. J. (1970), 'Does TF really work?', *Harv. Bus. Rev.,* vol. 48, no. 6, November–December, pp. 16–28.

Dosi, G. (1981), *Technical Change, Industrial Transformation and Public Policies: the Case of the Semi-conductor Industry,* Sussex European Research Centre, University of Sussex.

Dosi, G. (1982), 'Technical paradigms and technological trajectories—a suggested interpretation of the determinants and directions of technical change', *Research Policy,* vol. 11, no. 3.

Dow, J. C. R. (1964), *The Management of the British Economy 1945-1960*, NIESR and Cambridge University Press.

Downie, J. (1958), *The Competitive Process*, Duckworth.

Dubos, R. (1970), *Reason Awake: Science for Man*, Columbia University Press.

Dudintsev, V. (1957), *Not by Bread Alone*, Hutchinson.

Van Duijn, J. J. (1979), *De lange Golf in de Economie: kan Innovatie ons uit het dal Helpen?*, Assen, Van Gorcum.

Eads, G. and Nelson, R. R. (1971), 'Government support of advanced civilian technology', *Public Policy*, vol. 19, no. 3, pp. 405-27.

Encel, S. (1970), 'Science, discovery and innovation: an Australian case history', *Int. Soc. Sci. Journal*, vol. 22, no. 1, UNESCO.

Encel, S. *et al.* (1975), *The Art of Anticipation: Values and Methods in Forecasting*, Martin Robertson, London.

Enos, J. L. (1962a), *Petroleum Progress and Profits: a History of Process Innovation*, MIT Press.

Enos, J. L. (1962b), 'Invention and innovation in the petroleum refining industry', in National Bureau of Economic Research, *The Rate and Direction of Inventive Activity*, Princeton University Press.

Fabian, Y. (1963), *Measurement of Output of R and D*, OECD, DAS/PD/63.48.

Fabian, Y. (1965), 'Siderurgie et croissance économique en France et en Grande Bretagne (1735-1913)—les brevets en G-B', *Cahiers de l'ISEA*.

Federation of British Industries (1947), *Scientific and Technical Research in British Industry*.

Federation of British Industries (1961), *Industrial Research in Manufacturing Industry*.

Fels, R. (ed.) (1964), abridged edition of Schumpeter, J. A. (1939), *Business Cycles: a Theoretical, Historical and Statistical Analysis of the Capitalist Process*, McGraw Hill, New York.

Fisher, F. M., Griliches, Z., and Kaysen, C. (1962), 'The costs of automobile model changes since 1949', *J. Polit. Econ.*, vol. 70, no. 5, pp. 433-51.

Freeman, C. (1967), 'Science and economy at the national level', in OECD, *Problems in Science Policy*.

Freeman, C. (1971), *The Role of Small Firms in Innovation in the United Kingdom since 1945*, Report to the Bolton Committee of Inquiry on Small Firms, Research Report No. 6, HMSO.

Freeman, C. (1982a), 'Some economic implications of microelectronics', in Cohen, D. (ed.), *Agenda for Britain: Micro-Policy (Choices for the 80s)*, Philip Allen, Oxford, pp. 53-88.

Freeman, C. (ed.) (1982b), *Long Waves in the World Economy*, Butterworth, in press.

Freeman, C., Clark, J., and Soete, L. L. G. (1982), *Unemployment and Technical Innovation: a Study of Long Waves in Economic Development*, Frances Pinter, London.

Freeman, C., Cooper, C. M., and Pavitt, K. L. R. (1978), 'Policies for technical change', Chapter 7 in Freeman, C. and Jahoda, M. (eds), *World Futures: the Great Debate*, Martin Robertson.

Freeman, C., Curnow, R. C., Fuller, J. K., Robertson, A. B., and Whittaker, P. J. (1968), 'Chemical process plant: innovation and the world market', *Nat. Inst. Econ. Rev.*, no. 45.

Freeman, C., Harlow, C. J., and Fuller, J. K. (1965), 'Research and development in electronic capital goods', *Nat. Inst. Econ. Rev.*, no. 34.

Freeman, C. and Jahoda, M. (eds) (1968), *World Futures: The Great Debate*, Martin Robertson.

Freeman, C., Oldham, C. H. G., and Cooper, C. M. (1971), 'The goals of R and D in the 1970s', *Science Studies*, vol. 1, no. 3, pp. 357-406.

Freeman, C., Young, A., and Fuller, J. K. (1963), 'The plastics industry: a comparative study of research and innovation', *Nat. Inst. Econ. Rev.*, no. 26.

Freeman, C. and Young, A. (1965), *The Research and Development Effort in Western Europe, North America and the Soviet Union*, OECD.

Frowen, S. (ed.) (1982), *Controlling Industrial Economies*, Macmillan, London.

Gabor, D. (1964), *Inventing the Future*, Penguin.

Galbraith, J. K. (1969), *The New Industrial State*, Penguin.

Gartmann, H. (1959), *Sonst Stunde die Welt Still*, Dusseldorf.

Gazis, D. C. (1979), 'The influence of technology on science: some experience at IBM research', *Research Policy*, vol. 8, pp. 244-59.

Gelderen, J. van (1913), 'Springvloed: beschouwingen over industriele ontwikkeling en prijsbeweging', *De Nieuwe Tijd*, vol. 18, nos. 4, 5 and 6.

Gibbons, M. and Johnston, R. D. (1972), *The Interaction of Science and Technology*, Department of Liberal Studies in Science, University of Manchester, mimeo.

Gibbons, M. and Johnston, R. D. (1974), 'The roles of science in technological innovation', *Research Policy*, vol. 3, no. 3, pp. 220–42.

Gibbons, M. and Littler, D. (1979), 'The development of an innovation: the case of Porvair', *Research Policy*, vol. 8, no. 1, pp. 2–25.

Giersch, H. (1979), 'Aspects of growth, structural change and employment—a Schumpeterian perspective', *Weltwirtschaftliches Archiv*, vol. 115, no. 4, pp. 629–51.

Giersch, H. (ed.) (1982), *Proceedings of Conference on Emerging Technology at Kiel Institute of World Economics, 1981*, J. C. B. Mohr, Tübingen.

Gold, B. (1971), *Explorations in Managerial Economics*, Basic Books.

Gold, B. (1979), *Productivity, Technology and Capital*, D. C. Heath and Co., Lexington Books.

Gold, B. (1981), 'Technological diffusion in industry: research needs and shortcomings', *Journal of Industrial Economics*, March, pp. 247–69.

Golding, A. M. (1972), 'The Semi-conductor Industry in Britain and the United States: a Case Study in Innovation, Growth and the Diffusion of Technology', D.Phil. thesis, University of Sussex.

Granstrand, O. (1982), *Technology, Management and Markets*, Frances Pinter, London.

Greenberg, D. S. (1969), *The Politics of American Science*, Penguin.

Grosvenor, W. M. (1929), 'The seeds of progress', *Chemical Markets*.

Gruber, W. J. and Marquis, D. G. (1969), *Factors in the Transfer of Technology*, MIT.

Haber, L. F. (1958), *The Chemical Industry during the Nineteenth Century*, Oxford.

Haber, L. F. (1971), *The Chemical Industry, 1900-1930*, Oxford University Press.

Hamberg, D. (1964), 'Size of firm, oligopoly and research: the evidence', *Canadian J. Econ. Polit. Sci.*, vol. 30, no. 1, pp. 62–75.

Hamberg, D. (1966), *R and D: Essays in the Economics of Research and Development*, Random House.

Hardie, D. W. F. and Pratt, J. D. (1966), *A History of the Modern British Chemical Industry*, Pergamon.

Hart, A. (1966), 'A chart for evaluating product R and D projects', *Operational Research Quarterly*, vol. 17, no. 4, pp. 347–58.

Heath, J. B. (1971), *International Conference on Monopolies, Mergers and Restrictive Practices*, HMSO.

Helms, R. B. (ed.) (1981), *Drugs and Health: Economic Issues and Policy Objectives*, American Enterprise Institute for Public Policy Research.

Herrera, A. O. (1981), *La Larga Jornada: la Crisis Nuclear y el Destino Biológico del Hombre*, Siglo Veintiuno Editores.

Hessen, B. (1931), 'The social and economic roots of Newton's *Principia*', in N. Bukharin (ed.) (1971), *Science at the Cross-roads*, with an introduction by P. G. Werskey, Cass.

Hippel, E. von (1976), 'The dominant role of users in the scientific instrument innovation process', *Research Policy*, vol. 5, no. 3, pp. 212–39.

Hippel, E. von (1978), 'A customer-active paradigm for industrial product idea generation', *Research Policy*, vol. 7, no. 2, pp. 240–66.

Hirsch, S. (1965), 'The United States electronics industry in international trade', *Nat. Inst. Econ. Rev.*, no. 34.

Hitch, C. J. (1962), 'Comment' in National Bureau of Economic Research, *The Rate and Direction of Inventive Activity*, Princeton University Press.

Hoffmann, W. D. (1976), 'Market structure and strategies of R and D behaviour in the data-processing market', *Research Policy*, vol. 5, pp. 334–53.

Hollander, S. (1965), *The Sources of Increased Efficiency: a Study of Du Pont Rayon Plants*, MIT Press.

Hollingdale, S. H. and Toothill, G. C. (1965), *Electronic Computers*, Penguin.

Holroyd, Sir R. (1964), 'Productivity of industrial research with particular reference to research in chemical industry', in Institute of Chemical Engineers, *Proceedings of Symposium on Productivity in Research*, p. 6.

Hufbauer, G. C. (1966), *Synthetic Materials and the Theory of International Trade*, Duckworth.

Huxley, J. S. (1934), *Scientific Research and Social Needs*, Watts, London.

Illinois Institute of Technology Research Institute (1969), *Report on Project TRACES*, National Science Foundation.

Imperial Chemical Industries (1971), *Annual Report*.

Industrial Research Institute Research Corporation (1979), *Contribution of Basic Research to Recent Successful Industrial Innovations*, prepared for National Science Foundation, St. Louis: IRI/RC (PB 80-160179).

International Development Research Centre (1972), *Annual Report 1971-72*, Ottawa.
Irvine, J. H. and Martin, B. R. (1980), 'A methodology for assessing the scientific performance of research groups', *Scientia Yugoslavica*, vol. 6, nos. 1-4, pp. 83-95.
Irvine, J. H. and Martin, B. R. (1981), 'L'évaluation de la recherche fondamentale: est-elle possible?', *La Recherche*, no. 128, pp. 1406-16.
Jahoda, M. (1982), *The Social Psychology of Employment and Unemployment*, Cambridge University Press.
Jantsch, E. (1967), *Technological Forecasting in Perspective*, OECD.
Jervis, P. (1972), 'Innovation in electron-optical instruments', *Research Policy*, vol. 1, no. 2, p. 174.
Jewkes, J., Sawers, D., and Stillerman, R. (1958), *The Sources of Invention*, Macmillan (rev. edn. 1969).
Jones, D. T. (1981), 'Catching up with our competitors: the role of industrial policy' in Carter, C. (ed.), *Industrial Policy and Innovation*, pp. 146-56, Heinemann, London.
Jones, P. M. S. (1969), *Technological Forecasting as a Management Tool*, Programmes Analysis Unit, PAU M10.
Jones, R. V. (1978), *Most Secret War: British Scientific Intelligence 1939-1945*, Hamish Hamilton, London.
Jones, R. (ed.) (1981), *Readings from 'Futures'*, Westbury House, Guildford.
Kamien, M. I. and Schwartz, N. L. (1975), 'Market structure and innovation: a survey', *The Journal of Economic Literature*, vol. 23, no. 1.
Kapitza, P. L. (1966), in *Pravda*, 20 January.
Katz, B. G. and Phillips, A. (1982), 'Government, economies of scale and comparative advantage: the case of the computer industry', in Giersch, H. (ed.), *Proceedings of Conference on Emerging Technology, Kiel Institute of World Economics*, J. C. B. Mohr, Tübingen.
Kaufman, M. (1963), *First Century of Plastics*, Plastics Institute.
Kaufman, M. (1969), *The History of PVC: the Chemistry and Industrial Production of Polyvinyl Chloride*, Maclaren, London.
Kay, N. M. (1979), *The Innovating Firm: a Behavioural Theory of Corporate R & D*, Macmillan, London.
Keck, O. (1977), 'Fast breeder reactor development in West Germany: an analysis of government policy', D.Phil. thesis, University of Sussex.
Keck, O. (1980), 'Government policy and technical choice in the West German reactor programme', *Research Policy*, vol. 9, no. 4, pp. 302-56.
Keck, O. (1982), *Policy-making in a Nuclear Reactor Programme: the Case of the West German Fast Breeder Reactor*, Lexington Books, D. C. Heath and Co., Lexington.
Keirstead, B. S. (1948), *The Theory of Economic Change*, Macmillan, Toronto.
Kennedy, C. and Thirlwall, A. P. (1971), 'Surveys in applied economics: technical progress', *Economic Journal*, March.
Keynes, J. M. (1936), *General Theory of Employment, Interest and Money*, Macmillan.
Klein, B. H. (1977), *Dynamic Economics*, Harvard University Press, Cambridge, Mass.
Kleinknecht, A. (1981), 'Observations on the Schumpeterian swarming of innovations', *Futures*, vol. 13, no. 4, pp. 293-307.
Kleinman, H. S. (1975), *Indicators of the Output of New Technological Products from Industry*, report to US National Science Foundation, National Technical Information Service, US Department of Commerce.
Knight, F. H. (1965), *Risk, Uncertainty and Profit*, Harper.
Knox, F. (1969), *Consumers and the Economy*, Harrap.
Kondratiev, N. (1925), 'The major economic cycles', *Voprosy Konjunktury*, vol. 1, pp. 28-79; English translation reprinted in *Lloyd's Bank Review*, no. 129 (1978).
Krauch, H. (1970), *Prioritäten für die Forschungspolitik*, Carl Hanser Verlag, Munich.
Krauch, H. (1972), *Computer Demokratie*, VDI Verlag, Dusseldorf.
Kristensen, T. (1981), *Inflation and Unemployment in Modern Society*, Praeger, New York.
Kuznets, S. (1930), *Secular Movements in Production and Prices*, Houghton-Mifflin, Boston.
Kuznets, S. (1940), 'Schumpeter's business cycles', *American Economic Review*, vol. 30, no. 2, pp. 257-71.
Langrish, J. *et al.* (1972), *Wealth from Knowledge*, Macmillan, pp. 72-6.
Lawson, W. D., Lynch, C. A., and Richards, C. J. (1965), 'Corfam: research brings chemistry to footwear', *Research Management*, vol. 8, no. 1, pp. 5-26.
Lazarsfeld, P., Jahoda, M., and Zeisel, H. (1981), *Les Chomeurs de Marienthal*, les Éditions de Minuit, Paris; French translation of original German edition, 1932.

Lenin, V. I. (1917), *State and Revolution*, Martin Lawrence.

Liebermann, M. G. (1978), 'A literature citation study of science-technology coupling in electronics', *Proceedings of the IEEE*, vol. 66, no. 1, pp. 4–13.

Little, A. D. (1963), *Patterns and Problems of Technical Innovation in American Industry*, USGPO.

Litvak, I. A. and Maule, C. J. (1972), 'Managing the entrepreneurial enterprise', *Business Quarterly*, vol. 37, p. 47.

Lonnroth, M. and Walker, W. (1979), *The Viability of the Civil Nuclear Industry*, Rockefeller Foundation, New York; Royal Institute for International Affairs, London.

Machlin, D. J. (1973), 'The Economics of Technical Change in the Pottery Industry', M.A. dissertation, University of Keele.

Machlup, F. (1962a), *The Production and Distribution of Knowledge in the United States*, Princeton University Press.

Machlup, F. (1962b), 'The supply of inventors and inventions', in National Bureau of Economic Research, *The Rate and Direction of Inventive Activity*, Princeton University Press.

Mackenzie, N. and Mackenzie, J. (1973), *H. G. Wells: a Biography*, Simon and Schuster, New York.

Maclaurin, W. R. (1949), *Invention and Innovation in the Radio Industry*, Macmillan.

Mahdavi, K. B. (1972), *Technological Innovation: an Efficiency Investigation*, Beckmans, Stockholm.

Mandel, E. (1972), *Der Spätkapitalismus*, Suhrkampf, Frankfurt; English edition, *Late Capitalism* (1975), New Left Books, London.

Mandel, E. (1975), *Late Capitalism*, revised English edition of Mandel (1972), New Left Books, London.

Mandel, E. (1980), *Long Waves of Capitalist Development: the Marxist Interpretation*, Cambridge University Press.

Mansfield, E. (1961), 'Technical change and the rate of imitation', *Econometrics*, vol. 29, no. 4, pp. 741–66.

Mansfield, E. (1968a), *Industrial Research and Technological Innovation*, Norton.

Mansfield, E. (1968b), *The Economics of Technological Change*, Norton.

Mansfield, E. *et al.* (1971), *Research and Innovation in the Modern Corporation*, Norton.

Mansfield, E. *et al.* (1977), *The Production and Application of New Industrial Technology*, Norton.

Marquis, D. G. (1968), *Research Programme on the Management of Science and Technology*, MIT.

Marris, R. (1964), *The Economic Theory of Managerial Capitalism*, Macmillan.

Marschak, T., Glennan, T. K., and Summers, R. (1967), *Strategy for R and D*, Springer-Verlag.

Marshall, A. W. and Meckling, W. H. (1962), 'Predictability of the costs, time and success of development', in National Bureau of Economic Research, *The Rate and Direction of Inventive Activity*, Princeton University Press, pp. 461–77.

Martin, B. R. and Irvine, J. H. (1982), 'Assessing basic research: some partial indicators of scientific progress in radio astronomy', *Research Policy* (in press).

Marx, K. (1858), *Grundrisse*, Allen Lane edn., 1973.

Marx, K. and Engels, F. (1848), *Manifesto of the Communist Party*, London.

Matthews, R. (1968), 'Why has Britain had full employment since the war?', *Economic Journal*, vol. 78, pp. 555–69.

McCutcheon, R. (1972), 'High Flats in Britain, 1945–71', M.Sc. dissertation, University of Sussex.

McKay, A. L. and Bernal, J. D. (1966), 'Towards a science of science', *Technologist*, vol. 2, no. 4, pp. 319–28.

Meadows, D. *et al.* (1972), *The Limits to Growth*, Universe Books, New York.

Melto, de D. P. *et al.* (1980), *Innovation and Technological Change in Five Canadian Industries*, Discussion Paper no. 176, Economic Council of Canada, Ottawa.

Mensch, G. (1975), *Das technologische Patt: Innovation überwinden die Depression*, Umschau, Frankfurt; English edition: *Stalemate in Technology: Innovations overcome Depression*, Ballinger, New York, 1979.

Metcalfe, J. S. (1970), 'The diffusion of innovation in the Lancashire textile industry', *Manchester School of Economics and Social Studies*, no. 2, pp. 145–62.

Metcalfe, J. S. (1981), 'Impulse and diffusion in the study of technical change', *Futures*, vol. 13, no. 5, pp. 347–59.

Mishan, E. J. (1969), *The Costs of Economic Growth*, Penguin.

Morand, J. C. (1968), 'La recherche et la développement selon la dimension des enterprises', *Le Progrès Scientifique*, no. 122.

Morand, J. C. (1970), *Recherche et Dimension des Entreprises dans la Communauté Economique Européene*, Nancy.

Mowery, D. and Rosenberg, N. (1979), 'The influence of market demand upon innovation: a critical review of some recent empirical studies', *Research Policy*, vol. 8, pp. 102–53.

Mueller, D. C. (1966), 'Patents, research and development, and the measurement of inventive activity', *J. Indust. Econ.*, vol. 15, no. 1, pp. 26–37.

Mueller, W. F. (1962), 'The origins of the basic inventions underlying Du Pont's major product and process innovations, 1920–1950', in National Bureau of Economic Research, *The Rate and Direction of Inventive Activity*, Princeton University Press.

Musson, A. E. and Robinson, E. (1969), *Science and Technology in the Industrial Revolution*, Manchester University Press.

Naslund, B. and Sellstedt, B. (1972), *The Implementation and Use of Models for R and D Planning*, European Institute for Advanced Studies in Management.

Naslund, B. and Sellstedt, B. (1974), 'Budgets for research and development: an empirical study of 69 Swedish firms', *R and D Management*, vol. 4, pp. 67–73.

National Institute of Economic and Social Research (1963), 'Fast and slow-growing products in world trade', *National Institute Economic Review*, no. 25.

National Science Foundation (1961), *Publication of Basic Research Findings in Industry, 1957-1959*, NSF 61–62.

National Science Foundation (1973), *Interactions of Science and Technology in the Innovation Process*, Final Report from the Battelle Columbus Laboratory, NSF–667.

National Science Foundation (1981), *Science Resources Studies: Highlights*, NSF 81–331, Washington, D.C.

Neisser, H. P. (1942), 'Permanent technological unemployment', *American Economic Review*, vol. 32, no. 1, pp. 50–71.

Nelson, R. R. (1959), 'The simple economics of basic scientific research', *J. Polit. Econ.*, vol. 67, pp. 297–306.

Nelson, R. R. (1962), 'The link between science and invention: the case of the transistor', in National Bureau of Economic Research, *The Rate and Direction of Inventive Activity*, Princeton University Press.

Nelson, R. R. (1965), 'The allocation of R and D resources', in R. A. Tybout (ed.), *The Economics of R and D*, Ohio.

Nelson, R. R. (1971), *Issues and Suggestions for the Study of Industrial Organization in a Regime of Rapid Technical Change*, Yale University Economic Growth Centre, Discussion Paper no. 103.

Nelson, R. R. (1977), *The Moon and the Ghetto: an Essay on Public Policy Analysis*, Norton, New York.

Nelson, R. R. (1980), 'Parsimony, responsiveness and innovativeness as virtues of private enterprise: an exegesis of tangled doctrine', mimeo, Institute for Social and Policy Studies, Yale.

Nelson, R. R., Peck, J., and Kalachek, E. (1967), *Technology, Economic Growth and Public Policy*, Brookings Institution.

Nelson, R. R. and Winter, S. G. (1977), 'In search of useful theory of innovation', *Research Policy*, vol. 6, pp. 36–76.

Nelson, R. R. and Winter, S. G. (1982), 'The Schumpeterian trade-off revisited', *American Economic Review*, vol. 72, no. 1, March, pp. 114–33.

Nevins, A. and Hill, F. (1954), *Ford*, 2 vols., Scribner's, New York.

Norris, K. P. (1971), 'The accuracy of project cost and duration estimates in industrial R and D', *R and D Management*, vol. 2, no. 1, pp. 25–36.

OECD Study Group in the Economics of Education (1964), *The Residual Factor and Economic Growth*, Organization for Economic Cooperation and Development, Paris.

OECD (1966), *Government and Allocation of Resources to Science*, Part III, Organization for Economic Cooperation and Development, Paris.

OECD (1967), *The Overall Level and Structure of R and D Efforts in Member Countries*, Organization for Economic Cooperation and Development, Paris.

OECD (1968), *Gaps in Technology: Electronic Components*, Organization for Economic Cooperation and Development, Paris.

OECD (1971a), *R and D in OECD Member Countries: Trends and Objectives*, Organization for Economic Cooperation and Development, Paris.

OECD (1971b), *Science, Growth and Society* (Brooks Report), Organization for Economic Cooperation and Development, Paris.

OECD (1980a), *Technical Change and Economic Policy: Science and Technology in the New Economic Context*, Organization for Economic Cooperation and Development, Paris.

OECD (1980b), 'The measurement of the output of R and D activities—the 1980 conference on science and technology indicators', mimeo, Organization for Economic Cooperation and Development, Paris. (See also: OECD (1982), 'Workshop on patent and innovation statistics', mimeo).

OECD (1981a), *Science and Technology Indicators: Trends in Science and Technology in the OECD Area in the 1970s*, Organization for Economic Cooperation and Development, Paris.

OECD (1981b), *The Measurement of Scientific and Technical Activities: Proposed Standard Practice for Surveys of Research and Experimental Development* (Frascati Manual), Organization for Economic Cooperation and Development, Paris.

Olin, J. (1972), *R and D Management Practices: Chemical Industry in Europe*, Stanford Research Institute, Zurich.

Pareto, V. (1913), 'Alcuni relazioni fra la stato sociale e la variazoni della prosperita economica', *Rivista Italiana di Sociologia*, September–December, pp. 501–48.

Pavitt, K. L. R. (1971), *The Conditions for Success in Technological Innovation*, OECD, Paris.

Pavitt, K. L. R. (1980), *Technical Innovation and British Economic Performance*, Macmillan, London.

Pavitt, K. L. R. (1981), 'Technology in British industry: a suitable case for improvement', in Carter, C. (ed.), *Industrial Policy and Innovation*, pp. 88–115, Heinemann, London.

Pavitt, K. L. R. (1982), 'R and D, patenting and innovative activities: a statistical exploration', *Research Policy*, vol. 11, no. 1, pp. 35–51.

Pavitt, K. L. R. and Soete, L. L. G. (1980), 'Innovative activities and export shares', in Pavitt, K. (ed.), *Technical Innovation and British Economic Performance*, Macmillan, London.

Pavitt, K. L. R. and Walker, W. (1976), 'Government policies towards industrial innovation: a review', *Research Policy*, vol. 5, no. 1, pp. 11–97.

Payne, A. R. and Whittaker, R. E. (1972), 'Substitute materials for leather', *Materials Science and Engineering*, vol. 10, pp. 189–93.

Peck, J. (1968), 'British science and technology: the costs of over-commitment and gains from selectivity', in R. Caves (ed.), *Britain's Economic Prospects,* chapter 10, Brookings Institute.

Peck, J. and Scherer, F. M. (1962), *The Weapons Acquisitions Process: an Economic Analysis*, Harvard University Press.

Peck, M. J. and Wilson, R. (1982), 'Innovation, imitation and comparative advantage: the case of the consumer electronics industry', in Giersch, H. (1982).

Penrose, E. T. (1959), *The Theory of the Growth of the Firm*, Blackwell.

Petrov, V. M. (1967), *Ekonomicheskie Problemy Sodvuzhestva Nauki i Proizvodstra*, Leningrad.

Phillips, A. (1971), *Technology and Market Structure*, Lexington.

Phillips, A. (1980), 'Organisational factors in R and D and technological changes: market failure considerations', in Sahal, D. (ed), *Research, Development and Technological Innovation*, Lexington.

Plowden Report (1965), *Report of the Committee of Inquiry into the Aircraft Industry*, Cmnd. 2853, HMSO.

Polanyi, M. (1962), 'The Republic of Science', *Minerva*, vol. 1, no. 1, pp. 54–72.

Poole, J. B. and Andrews, K. (1972), *The Government of Science in Britain*, Weidenfeld & Nicolson.

Porat, M. U. (1977), *The Information Economy: Definition and Measurement*, vols. 1–9, USGPO, Washington.

Posner, M. (1961), 'International trade and technical change', *Oxford Econ. Papers*, vol. 13, no. 3, October, pp. 323–41.

Postan, M. M., Hay, D., and Scott, J. D. (1964), 'Design and development of weapons', in *History of Second World War*, HMSO.

Price, D. J. de Solla (1963), *Little Science, Big Science*, Columbia University Press.

Price, D. J. de Solla (1965), 'Is technology historically independent of science?', *Technology and Culture*, vol. VI, no. 4, p. 553.

Price, D. J. de Solla (1967), 'Research on research' in D. L. Arm (ed.), *Journeys in Science: Small Steps, Great Strides*, University of New Mexico Press.

Price, W. J. and Bass, L. W. (1969), 'Scientific research and the innovative process', *Science*, vol. 164, no. 3881, pp. 802–6.

Priest, W. C. and Hill, C. T. (1980), *Identifying and Assessing Discrete Technological Innovations*, National Science Foundation, Division of Science Resources Studies.

Radio Corporation of America (1963), *Three Historical Views*.

Ray, G. F. (1980), 'Innovation in the long cycle', *Lloyd's Bank Review*, no. 135, pp. 14–28.

Report of the Committee on Corrosion and Protection (1971), Department of Trade and Industry, HMSO, London, p. 128.

Roberts, E. B. (1968), 'The myths of research management', *Science and Technology*, no. 80, pp. 40-6.

Robertson, A. B., and Frost, M. (1978), 'Duopoly in the scientific instrument industry: the milk analyser case', *Research Policy*, vol. 7, no. 3, pp. 292–316.

Rogers, E. M. (1962), *Diffusion of Innovations*, Free Press, New York.

Rosenberg, N. (1976), *Perspectives on Technology*, Cambridge University Press.

Rothschild Report (1971), *A Framework for Government Research and Development*, Cmnd. 4814, HMSO.

Rothwell, R. *et al.* (1974), 'SAPPHO updated—Project SAPPHO phase 2', *Research Policy*, vol. 3, no. 3, pp. 258–91.

Rothwell, R. (1976), *Innovation in Textile Machinery: Some Significant Factors in Success and Failure*, SPRU Occasional Paper No. 2, University of Sussex.

Rothwell, R. (1979), *Technical Change and Competitiveness in Agricultural Engineering: the Performance of the UK Industry*, SPRU Occasional Paper No. 9, University of Sussex.

Rothwell, R. and Zegveld, W. (1981), *Industrial Innovation and Public Policy: Preparing for the 1980s and 1990s*, Frances Pinter, London.

Rothwell, R. and Zegveld, W. (1982), *Innovation and the Small and Medium-sized Firm*, Frances Pinter, London.

Rubenstein, A. (1966), 'Economic evaluation of R and D: a brief survey of theory and practice', *J. Indust. Eng.*, vol. 17, no. 11, pp. 615–20.

Rush, H. J., MacKerron, G. S., and Surrey, A. J. (1977), 'The advanced gas-cooled reactor: a case-study in reactor choice', *Energy Policy*, vol. 5, no. 2, pp. 95–105.

Saechtling, H. (1961), *Werkstoffe aus Menschenhand*, Munich.

Samuelson, P. A. (1967), *Economics*, 7th edn., McGraw-Hill.

Scherer, F. M. (1965a), 'Size of firm, oligopoly and research: a comment', *Canadian J. Econ. Polit. Sci.*, vol. 31, no. 2, pp. 256–66.

Scherer, F. M. (1965b), 'Firm size, market structure, opportunity and the output of patented inventions', *Amer. Econ. Rev.*, pp. 1097–123.

Scherer, F. M. (1980), *Industrial Market Structure and Economic Performance*, second edition, Rand McNally, Chicago.

Schmookler, J. (1966), *Invention and Economic Growth*, Harvard University Press.

Schott, B. and Graebner, K. (1974), 'R and D, innovation and micro-economic growth: a case study', *Research Policy*, vol. 2, no. 4, pp. 380–403.

Schott, B. and Muller, W. (1975), 'Process innovations and improvements as a determinant of the competitive position in the international plastic market', *Research Policy*, vol. 4, pp. 88–105.

Schott, K. (1975), 'Industrial R and D expenditures in the UK: an econometric analysis', D.Phil. thesis, Oxford University.

Schott, K. (1976), 'Investment in private industrial R and D in Britain', *Journal of Industrial Economics*, vol. 25, no. 2, pp. 81–99.

Schott, K. (1978), 'The relations between industrial R and D and factor demands', *Economic Journal*, vol. 88, March, pp. 85–106.

Schumacher, E. F. (1973), *Small is Beautiful: a Study of Economics as if People Mattered*, Bland & Briggs.

Schumpeter, J. A. (1912), *Theorie der wirtschaftlichen Entwicklung*, Leipzig, Duncker & Humboldt; English translation, *The Theory of Economic Development*, Harvard, 1934.

Schumpeter, J. A. (1928), 'The instability of capitalism', *Economic Journal*, pp. 361–86.

Schumpeter, J. A. (1939), *Business Cycles: a Theoretical, Historical and Statistical Analysis of the Capitalist Process*, 2 vols., McGraw-Hill, New York.

Schumpeter, J. A. (1942), *Capitalism, Socialism and Democracy*, Harper & Row.

Sciberras, E. (1977), *Multinational Electronic Companies and National Economic Policies*, JAI Press, Greenwich.

Sciberras, E. (1980), 'Technical innovation and international competitiveness in the television industry', mimeo, Science Policy Research Unit, University of Sussex.

Sciberras, E., Swords-Isherwood, N., and Senker, P. (1978), *Competition, Technical Change and Manpower in Electronic Capital Equipment: a Study of the UK Mini-computer Industry*, SPRU Occasional Paper No. 8, University of Sussex.

Science Policy Research Unit (1972), *Success and Failure in Industrial Innovation*, Centre for the Study of Industrial Innovation, London.

Seiler, R. (1965), *Improving the Effectiveness of Research and Development*, McGraw-Hill.

Senker, P. J. and Swords-Isherwood, N. B. (1980), *Microelectronics and the Engineering Industry: the Need for Skills*, Frances Pinter, London.

Sercovich, F. (1974), 'Foreign Technology and Control in Argentinian Industry', D.Phil. thesis, University of Sussex.

Servan-Schreiber, J. (1965), *The American Challenge*, Penguin, Harmondsworth (English translation of *Le Défi Américain*, Paris, 1965).

Shackle, G. L. S. (1955), *Uncertainty in Economics and other Reflections*, Cambridge University Press.

Shackle, G. L. S. (1961), *Decision, Order and Time in Human Affairs*, Cambridge University Press.

Sherwin, C. W. and Isenson, R. S. (1966), *First Interim Report on Project 'Hindsight'*, Office of the Director of Defence Research and Engineering, Washington.

Shimshoni, D. (1966), 'Aspects of Scientific Entrepreneurship', D.Phil. thesis, Harvard.

Shimshoni, D. (1970), 'The mobile scientist in the American instrument industry', *Minerva*, vol. 8, no. 1, pp. 59–89.

Silk, L. S. (1960), *The Research Revolution*, McGraw-Hill.

Sirilli, G. (1982), 'The researcher in Italian industry', mimeo, Science Policy Research Unit, University of Sussex.

Smith, A. (1776), *An Inquiry into the Nature and Causes of the Wealth of Nations*, Dent edn. (1910), p. 8.

Soete, L. L. G. (1979), 'Firm size and inventive activity: the evidence reconsidered', *European Economic Review*, vol. 12, pp. 319–40.

Soete, L. L. G. (1981), 'A general test of technological gap trade theory', *Weltwirtschaftliches Archiv*, vol. 117, no. 4, pp. 638–66.

Solo, C. S. (1951), 'Innovation in the capitalist process: a critique of the Schumpeterian theory', *Q. J. Econ.*, vol. 65, no. 3, pp. 417–28.

Solo, R. A. (1961), *Patent Practices of the Department of Defence*, Committee on Judiciary, 72757, USGPO.

Solo, R. A. (1966), 'Patent policy for government-sponsored R and D', *Idea*, vol. 10, no. 2, pp. 143–206.

Solo, R. A. (1967), *Economic Organisation and Social Systems*, Bobbs-Merrill Inc., Indiana.

Solo, R. A. (1980), *Across the High Technology Threshold: the Case of Synthetic Rubber*, Norwood Editions, Norwood, Pa.

Spiegel-Rösing, I. and Price, D. de Solla (eds) (1977), *Science, Technology and Society: a Cross-disciplinary Perspective*, Sage, London and Beverley Hills.

Stoneman, P. (1976), *Technological Diffusion and the Computer Revolution: the UK Experience*, University of Cambridge, Department of Applied Economics Monographs, no. 25, Cambridge University Press.

Sturmey, S. G. (1958), *The Economic Development of Radio*, Duckworth.

Sunday Times (1970), *History of Inventions*, no. 9.

Surrey, A. J. and Chesshire, J. H. (1972), *The World Market for Electric Power Equipment*, Science Policy Research Unit, University of Sussex.

Surrey, A. J. and Thomas, S. D. (1980), *Worldwide Nuclear Plant Performance: Lessons for Technology Policy*, SPRU Occasional Paper No. 10, University of Sussex.

Szakasits, G. (1974), 'The adoption of the SAPPHO method in the Hungarian electronics industry', *Research Policy*, vol. 3, no. 1, pp. 18–28.

Taylor, C. and Silberston, A. (1973), *The Economic Impact of the Patent System*, Cambridge University Press.

Telefunken (1928), *25 Jahre Telefunken*.

Telefunken (1953), *50 Jahre Telefunken*.

Terleckyj, N. (1974), *The Effects of R and D on Productivity Growth in Industry*, NPA, Washington.

Ter Meer, F. (1953), *Die I G Farben*, Dusseldorf.

Teubal, M. *et al.* (1976), 'Performance in innovation in the Israeli electronics industry', *Research Policy*, vol. 5, no. 4, pp. 354–79.

Thomas, H. (1970), 'Econometric and Decisions Analysis: Studies in R and D in the Electronics Industry', Ph.D. thesis, University of Edinburgh.

Tilton, J. (1971), *International Diffusion of Technology: the Case of Semi-Conductors*, Brookings Institution.

Tisdell, C. A. (1981), *Science and Technology Policy: Priorities of Governments*, Chapman & Hall, London.

Townsend, E. C. (1969), *Investment and Uncertainty*, Oliver & Boyd.

Townsend, J. (1976), *Innovation in Coal-mining machinery—the Anderton Shearer-loader and the Role of the NCB and Supply Industry in its Development*, SPRU Occasional Paper No. 3, University of Sussex.

Townsend, J. *et al.* (1982), *Innovations in Britain since 1945*, SPRU Occasional Paper No. 16, University of Sussex.

Turner, D. F. and Williamson, O. E. (1969), 'Market structure in relation to technical and organizational innovation', in J. B. Heath (ed.), *International Conference on Monopolies, Mergers and Restrictive Practices*, HMSO, 1971.

UNESCO (1969), *The Measurement of Scientific and Technological Activities*, Paris.

UNESCO (1970), *Measurement of Output of Research and Experimental Development*, Paris.

United Kingdom Atomic Energy Authority (1971), *Seventeenth Annual Report, 1970-71*, HMSO.

United Nations (1970), *Science and Technology for Development*, Annex II.

United Nations Economic Commission for Europe (1968), *Policies and Means of Promoting Technical Progress*, New York.

United States Department of Commerce (1963), *Patterns and Problems of Technical Innovation in American Industry*, Report to National Science Foundation.

United States Department of Commerce (1967), *Technological Innovation: its Environment and Management*, USGPO.

Vernon, R. (1966), 'International investment and international trade in the product cycle', *Q. J. Econ.*, vol. 80, pp. 190–207.

Walsh, V., Townsend, J., Achilladelis, B. G., and Freeman, C. (1979), 'Trends in invention and innovation in the chemical industry', mimeo, Science Policy Research Unit, University of Sussex.

Ward, W. H. (1967), 'The sailing ship effect', *Bull. Inst. Physics*, vol. 18, p. 169.

Watson, T. J. (1963), *Meeting the Challenge of Growth*, McKinsey Foundation Lecture, no. 2.

Whiston, T. G. (1980), 'Product life for the automobile: a possible framework for analysing policy options', mimeo, Science Policy Research Unit, University of Sussex.

Wilkins, G. J. (1967), 'A record of innovation and exports', in G. Teeling-Smith (ed.), *Innovation and the Balance of Payments*, Office of Health Economics.

Wise, T. A. (1966), 'IBM's $5,000,000,000 gamble', *Fortune*, September.

Wynn, M. R. and Rutherford, G. H. (1964), 'Ethylene's unlimited horizons', *European Chemical News Large Plants Supplement*, 16 October.

Yarsley, V. E. and Couzens, E. G. (1956), *Plastics in the Service of Man*, Penguin.

Zuse (1961), *25 Jahre Entwicklung Programmgesteuerter Rechenanlagen*, Hersfeld.

Zwan, A. van der (1979), 'On the assessment of the Kondratiev cycle and related issues', Centre for Research in Business Economics, Rotterdam, mimeo.

Bibliographies, annotated bibliographies, and articles or books which contain useful reviews of earlier literature relating to various aspects of innovation

Blaug, M. (1963), 'A survey of the theory of process innovation', *Economica*, vol. 30, no. 1, pp. 13–32.

Freeman, C. and Pavitt, K. (eds) (1982), 'Master author index and master subject index to volumes 1–10 (1972–1982) of *Research Policy*', *Research Policy*, vol. 11, no. 1, pp. 57–82.

Kamien, M. and Schwartz, N. (1975), 'Market structure and innovation: a survey', *Journal of Economic Literature*, vol. 13, no. 1.

Kennedy, C. and Thirlwall, A. P. (1972), 'Technical Progress', in *Surveys of Applied Economics*, vol. 1, pp. 115–77, Royal Economic Society and Social Science Research Council, Macmillan, London.

Metcalfe, J. S. (1981), 'Impulse and diffusion in the study of technological change', *Futures*, vol. 13, no. 5, pp. 347–59.

Metcalfe, J. S. and Stubbs, P. C. (1980), 'Technical progress: a bibliography', in Puu, T. and Wibe, S. (eds), *The Economics of Technological Progress*, Macmillan.

Metcalfe, J. S. and Stubbs, P. C. (1982), *The Economics of Technical Change: an Annotated Bibliography*, Manchester University Press, forthcoming.

Mowery, D. and Rosenberg, N. (1979), 'The influence of market demand upon innovation: a critical review of some recent empirical studies', *Research Policy*, vol. 8, no. 2, pp. 102–50.

Nelson, R. R. and Winter, S. G. (1977), 'In search of useful theory of innovation', *Research Policy*, vol. 6, no. 1, pp. 37–76.

Pavitt, K. and Walker, W. (1976), 'Government policies towards industrial innovation: a review', *Research Policy*, vol. 5, no. 1, pp. 11–97.

Science Policy Research Unit (1983), *Bibliography on Technical Innovation*, Harvester Press, forthcoming.

Spiegel-Rösing, I. and Price, D. de Solla (eds) (1977), *Science, Technology and Society* (contains review chapters on sociology, economics, politics and psychology of science and on many aspects of policy for technology), Sage, London and Beverley Hills.

Index